A Treatise on
Time and Space

A Treatise on
Time and Space

J. R. LUCAS

Methuen & Co Ltd

First published 1973
by Methuen & Co Ltd
11 New Fetter Lane, London EC4
© 1973 J. R. Lucas
Printed in Great Britain by
William Clowes & Sons Limited
London, Colchester and Beccles

SBN 416 75070 2

Distributed in the U.S.A. by
Harper & Row Publishers Inc.
Barnes & Noble Import Division

To R. H. L.

Come, let us go, while we are in our prime,
And take the harmless folly of the time.
    We shall grow old apace, and die
    Before we know our liberty.
    Our life is short, and our days run
    As fast away as does the sun;
And, as a vapour or a drop of rain,
Once lost, can ne'er be found again,
    So when or you or I are made
    A fable, song, or fleeting shade,
    All love, all liking, all delight
    Lies drowned with us in endless night.
Then while time serves, and we are but decaying
Come, my Corinna, come, let's go a-Maying.

Robert Herrick

Apology

'Μηδεὶς ἀγεωμέτρητος εἰσίτω' Plato had written up over the entrance to the Academy; and any philosopher concerned with time and space will re-echo his words. It is impossible to treat these concepts adequately without mathematics. Mathematicians have developed the concepts we need if we are to answer the questions that common sense has posed; and have, in turn, raised further questions that can neither be answered nor even be asked without the aid of mathematics. Therefore I must use mathematics; and although I try to make it as simple and perspicuous as possible, and offer my apologies if I have gone into needless detail, I cannot dispense with formulae and equations altogether.

But, although I use mathematics, I am not offering a mathematical tract. Mathematics, partly as the result again of Plato's influence, has become a formal axiomatic discipline, with its own criteria of validity and elegance. I shall seldom put forward mathematical proofs. I shall indicate, non-deductively, a line of argument rather than deploy coercive deductive arguments. The reader will be able to pick holes in my arguments – but also, I hope, see how the holes could be blocked up again. There are many mathematically interesting possibilities that I do not discuss – that space might be multiply-connected, for instance. But I am concerned not with the formal possibilities of abstract spaces but with the actual features of our space. I may be wrong in what I take those features to be. But if I am wrong, it will be on other grounds than those of mathematical elegance.

Equally, the physics is both by choice and of necessity thin. I am not a physicist. I am not relating the latest discoveries physicists have made, nor am I joining the long queue of those who have sought to make relativity readable. Even if I could, I should not want to. For my intention is primarily philosophical, and although the philosopher has much indeed to learn from the physicist, his aims are different. He is trying to analyse, and possibly to revise, the structure of our concepts, rather than discover new true universal propositions about natural phenomena. The two are related. There is no merit in the philosopher arguing for views that are, as a matter of brute empirical fact, false; and physicists are often guided in their theorizing – sometimes more than they realize – by philosophical, or at least *a priori*, rather than purely empirical, considerations. But the emphasis is different. The philosopher is concerned with general, and often rather vague, concepts such as change, communication, sameness, difference, identity, substance; whereas the physicist is likely to refer us to much more specialized and definite concepts, such as gravitation, electricity, magnetism, neutrons, neutrinos and mesons. And of these, and of the many true things that can be said about them, I shall have little to tell.

Contents

I Time by itself

§ 1
The nature of time

Time is more fundamental than space. Indeed, time is the most pervasive of all the categories. Some theologians say that God is outside time, but it cannot be true of any personal God that he is timeless, for a personal God is conscious, and time is a concomitant of consciousness. Time is not only the concomitant of consciousness, but the process of actualization and the dimension of change. The many different definitions of time given by philosophers reflect its many different connections with other fundamental categories. Time is connected with persons, both as sentient beings and as agents; it is connected with modality, and the passage from the open future to the unalterable past; it is connected with change, and therefore with the things that change and the space in which they change.

These different connections are responsible both for the importance of the concept of time and for the difficulty we experience in defining it. We cannot philosophize long about any subject without coming up against the problem of time. But as we wrestle with it, we find ourselves repeating St Augustine's lament: *"Quid est ergo tempus? si nemo ex me quaerat, scio; si quaerenti explicare velim, nescio."* ("What is time? if nobody asks me, I know, but if I want to explain it to some one, then I do not know.") [1] It is a universal

[1]   St Augustine, *Confessions*, bk XI, ch. XIV, xvii.

experience. We know what time is, we are all familiar with it, but all feel uneasy when it comes to expressing it in words. We listen with respect when we are told that time is a moving image of eternity, εἰκὼ κινητόν τινα αἰῶνος [2], or the measure of change with respect to earlier and later, ἀριθμός κινήσεως κατὰ τὸ πρότερον καὶ ὕστερον [3], or that time is extension, extension perhaps of mind [4], or that it is a concomitant of concomitants, σύμπτωμα συμπτωμάτων [5], or that it is the order of events, or that it is the form of becoming, or the possibility of change, or, more modernly, that time is what the clocks say, or that it is the independent variable in the laws of mechanics [6], or that it is the fourth dimension, and therefore unreal and only imaginary, because it involves $\sqrt{-1}$.

But although we listen with respect, we do not give our wholehearted assent. None of these sayings, true, perhaps profound, though they are, seems to express all that we mean by the word 'time'. Time means more than all these. Not only do we use the word in many different locutions [7], but it seems something too universal, too pervasive, ever to be encapsulated in a few words. We cannot say what time is, because we know already, and our saying could never match up to all that we already know. Yet we cannot be content merely to know wordlessly, and to ward off the questioner, like Meno, by saying that he could not ask us what time was unless he already knew. For we get puzzled by time. Not only are we unable to answer other men's questions, but we cannot think straight about it by ourselves. We are always being driven into quandaries and paradoxes by the various analogies that present themselves. "Time like an ever-rolling stream bears all its sons away", but where to? The present is with us – we know that. But the future, which is not yet present with us, and the past, which no longer is present with us, where are they? Where are the days of yesteryear? Indeed, does even the present exist? How large is it? Surely, on reflection, we are compelled to admit that it is no size at all – but then it cannot constitute any part of time and must be altogether nothing. Or again, we talk of time going fast or going slow – "Time goes faster", we are told, "as one gets older"; but this, however natural, is also awkward, since it is only by reference to time that we measure fastness or slowness.

Measuring time seems equally mysterious. We can measure space by moving a ruler – a standard length – around, and, if in doubt, bringing it back again, and checking that all the readings are repeatable. But we cannot take a standard second around, and lay it off sixty times against a minute yesterday, and then

[2]   Plato, *Timaeus*, 38; see further below § 13, pp. 75–6
[3]   Aristotle, *Physics*, IV, 11, 220a 25.
[4]   St Augustine, *Confessions*, bk XI, ch. XXVI, xxxiii; quoted below § 2, p. 14.
[5]   Epicurus, as interpreted by Demetrius of Sparta *apud* Sextus Empiricus, *Adversus Mathematicos*, X (i.e. Πρὸς φυσικούς, B), 219.
[6]   Quoted, e.g. by H. Margenau, *The Nature of Physical Reality* (New York, 1950), p. 136.
[7]   For many other uses of the word 'time', see F. Waismann, "Analytic–Synthetic, II", *Analysis*, IX (December 1950), pp. 26–8.

against a minute today, and back to check yesterday's minute again. Measurements, so called, of time are essentially unrepeatable, and for that reason alone suspect to the scientific mind. Yet we rely on them with great confidence, so much so that we talk of what our measuring instruments read as being "the time". Perhaps we should not. A B.B.C. announcer, more refined, said "The right time is twenty to eight", but Miss P. Wacholder of Hull objected, "Surely *time* is never wrong – our clocks and watches may be, but never time." [8] It is difficult to talk, difficult to think, about time straight. It is too immediate and pervasive to get into focus, too intangible and insubstantial to grasp or comprehend.

Some of these difficulties are susceptible to straightforward philosophical analysis. By considering carefully how we know that some temporal assertion is correct or incorrect, and how we actually set about assigning magnitudes to temporal intervals, we can resolve many of our difficulties. But not all. Our difficulties about time are not due merely to our being insufficiently self-conscious about our use of language and measuring instruments. Time is puzzling not only because it is unlike anything else in our conceptual universe, but just because it is part of our conceptual structure and is connected to a number of other parts. These different connections give us different perspectives on the concept of time, indeed almost different concepts. Time is related to change, and through change to the things that change and the space in which they change. But quite apart from change, time is related to consciousness, and hence with persons. More important to our concept of personality than bare sentience is the fact that we are rational agents who make plans for the future and choose between alternatives presented to us and know ourselves as being the beings who have done what we have done. Time, therefore, is linked with personality in two ways: not only is it given to each one of us in experience, as an essential concomitant of consciousness, but it is a necessary condition of activity, and an integral part of each man's notion of what it is to be himself. Without time no agent could act, for to act is to bring about something that we want to come to pass, and time is the passage from possibility to actuality, from aspiration to achievement. If I am never able to give an answer to the question "What are you doing?" I am not a responsible agent, but insane. But if I can say what it is I am doing, then I must be able also to say what I have just been doing in the immediate past, and what I am just going to do in the immediate future; for actions are not isolated instantaneous happenings but are intelligible only within a pattern of activity. In order to be an agent, I must know what I am doing, I must remember what I have done in the immediate past and realize what I intend to do in the immediate future. Without time there could be no activity. And even if we could isolate individual actions it would still make no sense to talk of agents acting timelessly or of deeds done outside them.

[8]   *Radio Times* (11 February 1965).

Different approaches bring different aspects of time into prominence but no single approach can reveal the whole nature of time, and each needs to be supplemented on occasion by the others. In the same way there are many concepts of space, which in turn reflect their multiplicity back into time. It is for this reason that profound thinkers have given their different definitions of time, each one of which has seemed to reveal some important aspect of the truth. It may be possible in the end to give a unified account which reconciles their divergent insights in one coherent theory; but it would be a mistake to begin like this. Both the difficulty and the importance of the concept of time arise from there being different perspectives on it, which partly although not altogether overlap, and which we superimpose on one another to obtain a concept, or family of concepts, of great complexity and richness. To appreciate it fully, we need to approach it in different ways, and to experience the tension engendered by its diverse conceptual limbs. In particular, illumination often comes by considering the various puzzles that have perplexed men when they have thought about time. We shall consider first the puzzles that arise from our bare awareness of the passage of time. We shall start with the puzzle of time going quickly or slowly, which we shall resolve by distinguishing between public and private time; and then go on to deal with St Augustine's problem of the ever-shrinking present. These both turn on topological considerations. We shall complete a survey of the topological properties of time, which can largely be established by arguments from consciousness alone. We shall then consider time as the condition of activity and as the actualization of the potential future into the fixed past, and discuss the directedness and the serial nature of time. Only then shall we turn to the time of the scientists, the dimension of change and physical process, which can be measured but is too featureless and insubstantial to grasp or comprehend. This will lead us to one concept of space, which in turn leads us on to the more general concept still of space-time. Theories of space, together with those of Newtonian mechanics and relativity, have a profound bearing on our understanding of time, and raise questions that sometimes require a radical reappraisal of our untutored institutions and sometimes need, rather, a rational reaffirmation of what we have always known to be true.

§ 2
Time and consciousness

Time is related to consciousness. I cannot conceive of a mind being conscious
of something about whom the question 'when?' does not arise, whereas there
are many states of consciousness for which the question 'where?' does not
naturally arise. I can feel anger or elation, or can recollect the past, or can try
to prove a mathematical theorem without having any spatial experience or any
idea of where I am located in space; whereas I cannot enjoy any state of
consciousness at all without having some idea of its being before or after some
other states of consciousness, and being something of which it must always be
intelligible to ask when I had it. I may be unable to answer – our sense of time
is often unreliable – but it is necessarily an intelligible question; whereas the
question 'Where were you angry?' or 'Where did you try to prove Taylor's
theorem?' is not a very natural question, and need not be even intelligible.
Disembodied minds are conceivable and would be non-spatial, whereas timeless minds seem quite inconceivable. We might express this by saying that minds
are only contingently located in space, but necessarily "located" in time.
Space and time are thus importantly different. Whereas we do not have to be
able to ask or answer the question where a mind is having an experience, it is
unintelligible to claim that we cannot say, and cannot even ask, when it is, or

was, or will be, having an experience. And, *per contra*, whereas I can conceive of a space that is totally disconnected with the space in which I happen to be, I cannot conceive of a time disconnected with my time; that is, there cannot be a time or any temporal event such that the question 'What is its relation to me?' cannot arise. Witness the phrase 'Once upon a time ...', which by blocking the question 'when?' conveys the implication that what is narrated never really happened (see further below, § 7, p. 37).

The connexion between consciousness and time is important not only in showing the difference between time and space, but in refuting certain positivist analyses of time in terms of change, and in establishing the continuity of time independently of any physical considerations about the nature of change or of space. One of the reasons why time is puzzling is that we are immediately and irresistibly aware of it, and therefore often find the rational reconstructions offered by philosophers intuitively wrong. Although consciousness is only one aspect of personality, and although the connexions between time and modality and time and change are also fundamental, nevertheless it is as conscious beings that we first become aware of time.

Time is given to each one of us in experience. If we are aware of anything we are aware of the passage of time. Even if we are not acting, in the sense of initiating changes in the external world, and are not observing any changes in the external world, we still in the wakeful watches of the night can sense the sleepless hours slipping by:

Δέδυκε μὲν ἀ σελάννα
καὶ Πληιάδες, μέσαι δὲ
νύκτες, παρὰ δ' ἔρχετ' ὤρα,
ἔγω δὲ μόνα κατεύδω.

The Moon is gone
    And the Pleiads set,
Midnight is nigh.
    Time passes on,
    And passes, yet
    Alone I lie. [1]

Newton's tutor argued the same point, that while time as measurable implies motion, time by itself does not:

Finge stellas omnes ab incunabilis suis fixas perstitisse; nihil inde quicquam tempori decessisset; tamdieu quies ista perdurasset, quamdieu motus hic effluxit. Prius, posterius, simul (quoad ortus rerum et interitus) etiam in illo tranquillo statu fuisset in se, potuisset a mente magis perfecta apprehendi.

Imagine all the stars to have remained fixed from their birth; nothing would have been lost to time; as long would that stillness have endured as has

[1]   *Oxford Book of Greek Verse*, No. 156.

continued the flow of this motion. Before, after, at the same time (as far as concerns the rise and disappearance of things), even in that tranquil state would have had their proper existence, and might by a more perfect mind have been perceived. [2]

This view has been much contested. Aristotle defined time as the measure of motion, ἀριθμός κινήσεως. Lucretius was quite definite:

> Tempus item per se non est, sed rebus ab ipsis
> consequitur sensus, . . .
> nec per se quenquam tempus sentire fatendum est
> semotum ab rerum motu placida que quiete;

Time by itself does not exist, but the sense of time follows from things themselves: ... nor should it be said that any one experiences time by itself when he is removed from the calm and peaceful motion of things; [3]

and many scientists since have taken a positivist attitude, and defined time as "what the clocks say", with the implication that if all processes were stilled, there would be no passage of time either. It is a natural error to take the measure for the meaning, much as tough-minded lawyers are liable to affirm that "the law is what the judges say it is": but it is an error. At its crudest, the argument is based on the verification theory of meaning, which maintains that the meaning of words is constituted by the criteria for their use. The logical positivists of the 1930s carried the programme through remorselessly, to the extent of making the frequency theory of probability true by definition, and maintaining that a man was not really in pain unless he was showing signs of it that would justify another man in saying of him that he was in pain. But although frequencies are often the best evidence of probabilities, and overt behaviour the only evidence of other men's states of mind, we do not believe that in making a probability judgement a man is really referring only to the evidence already available rather than to the event still in issue, nor do we regard it as a necessary truth of logic that a man cannot be in pain unless he shows it. And the crude empiricist argument for defining time as the measure of motion (or the aspect of process) or as the concomitant of concomitants (or symptom of symptoms) is equally invalid.

The whole point of a criterion is that it is a criterion for something – for something other than itself. We are interested in a man's flinching or grimacing because we take it as showing that he is suffering pain: if it emerges that it is not a sign of his being in pain – for example, if we discover it is just a tic – we cease to notice it. Similarly the whole point of having clocks or judges is to tell

[2] Isaac Barrow, *Lectiones Geometricae* (London, 1970), *Lectio* 1, p. 3; reprinted in *Mathematical Works of Isaac Barrow, D.D.*, ed. W. Whewell (Cambridge, 1860), II, 161; translated and quoted by E. A. Burtt, *The Metaphysical Foundations of Modern Physical Science* (2nd ed. London, 1932), p. 150.
[3] *De Rerum Natura*, I, 459–60, 462–3.

the time or to tell the law. This is what clocks and judges are *for*. But that is intelligible only if we know without having a clock or a judge to hand what time or law roughly is [4]. Time is not what the clocks say, but what they are trying to tell, are there to tell.

We can argue more directly against the positivist thesis, by showing that it claims too much [5]. If time is to be defined in terms of change or motion or process, then Barrow's suggestion of an initial stationary state of the universe would be *logically* impossible. But it is not [6]. Even within the austere framework of Lemmon's minimal tense logic K_t [7], we can distinguish between a dawn of creation in which the stars started in their courses the moment time began and a more leisurely inauguration in which they spent part of the morning doing nothing in unison. In the former case, we can assert

$CPpPNPNq$, in Prior's Polish notation,
or
$Pp \supset P \sim P \sim q$, using the more familiar symbols for logical constants

⎫ Where P means 'it was sometime
⎬ the case that'
⎭

that is, if it was the case sometime that p, it was also the case sometime that it was not ever the case that q was false: and this is vacuously true at the beginning of time, because no statement in the past tense could then have been true, so that the antecedent would be false; and at all later dates the consequent would be true, for it was the case sometime – viz. at the beginning of time – that it was not ever before the case that q was false, because, again, at the beginning of time no statement in the past tense, not even the statement that q was false, could be true. But to say that time had a beginning is a stronger thesis than merely to maintain that before they began on their ceaseless motions the stars stayed still for a season. In that case we could not assert that, provided some statement, Pp, in the past tense could be made, we could say categorically of any q that it had been the case sometime that it was not ever the case that it was false: for there would be plenty of false propositions about the stationary stars – for example, that some of them were moving – which were far from never having been false. All that we should be entitled to assert would be the weaker thesis

$$CPpP\Pi qANPNqNPq$$
or
$$Pp \supset P[(\forall q)(\sim P \sim q \vee \sim Pq)]$$

[4]   Compare H. L. A. Hart, *The Concept of Law* (Oxford, 1961), p. 8; or O. W. Holmes, "The Path of the Law", in *Collected Papers* (New York, 1952, or 1920), p. 173. See also Plotinus, *Enneads*, III, 7, 9, esp. ll. 31–5.
[5]   The following argument is rather technical. The reader can resume the thread on p. 11.
[6]   I am indebted to the late A. N. Prior for drawing my attention to this argument, which is a development of one on p. 111 of his *Collected Papers on Time and Tense* (Oxford, 1948). He is dealing with the end of the universe and the end of time; but in this respect the beginning and the end are exactly similar.
[7]   See, e.g., A. N. Prior, *Past, Present and Future* (Oxford, 1967), p. 176.

that is, if it was the case sometime that p, then it was sometime also the case that every q either had never been false or had never been true; that is, provided any past statement at all can be made (i.e. excluding, if there is one, the very beginning of time), one can refer to some past time when there had never been any change, and any proposition either always had been true or always had been false. Such a thesis would hold at the beginning of time, since the antecedent would be false; and, at all later dates, if there had been an initial stationary state, since the consequent would then be true. Moreover, we can see that the stationary state thesis is weaker than the beginning of time one, and does not entail it, by interpreting all Prior's tense operators as unary null operators. The stationary state thesis then becomes the tautology

$$CpΠqAqNq$$
$$p \supset (\forall q)(q \lor \sim q)$$

which cannot together with the other tautologies of reinterpreted tense logic entail the non-tautological

$$Cpq$$
$$p \supset q$$

which is what the thesis that time has a beginning becomes. The two theses are logically distinct, and Barrow's stationary state is a logically possible one.

Aristotle's argument is different, but also invalid. He argues

'Ἀλλὰ μὴν οὐδ'ἄνευ γε μεταβολῆς· ὅταν γὰρ μηδὲν αὐτοὶ μεταβάλλωμεν τὴν διάνοιαν ἢ λάθωμεν μεταβάλλοντες, οὐ δοκεῖ ἡμῖν γεγονέναι χρόνος, καθάπερ οὐδὲ τοῖς ἐν Σαρδοῖ μυθολογουμένοις καθεύδειν παρὰ τοῖς ἥρωσιν, ὅταν ἐγερθῶσι· συνάπτουσι γὰρ τῷ πρότερον νῦν τὸ ὕστερον νῦν καὶ ἓν ποιοῦσιν, ἐξαιροῦντες διὰ τὴν ἀναισθησίαν τὸ μεταξύ. ὥσπερ οὖν εἰ μὴ ἦν ἕτερον τὸ νῦν ἀλλὰ ταὐτὸ καὶ ἕν, οὐκ ἂν ἦν χρόνος, οὕτως καὶ ἐπεὶ λανθάνει ἕτερον ὄν, οὐ δοκεῖ εἶναι τὸ μεταξὺ χρόνος. εἰ δὴ τὸ μὴ οἴεσθαι εἶναι χρόνον τότε συμβαίνει ἡμῖν, ὅταν μὴ ὁρίσωμεν μηδεμίαν μεταβολήν, ἀλλ' ἐν ἑνὶ καὶ ἀδιαιρέτῳ φαίνηται ἡ ψυχὴ μένειν, ὅταν δ' αἰσθώμεθα καὶ ὁρίσωμεν, τότε φαμὲν γεγονέναι χρόνον, φανερὸν ὅτι οὐκ ἔστιν ἄνευ κινήσεως καὶ μεταβολῆς χρόνος. [8]

[8] *Physics*, IV, 11, 218b 21–219a 1, tr. R. P. Hardie: "But neither does time exist without change; for when the state of our own minds does not change at all, or we have not noticed its changing, we do not realize that time has elapsed, any more than those who are fabled to sleep among the heroes in Sardinia do when they are awakened; for they connect the earlier 'now' with the later and make them one, cutting out the interval because of their failure to notice it. So, just as if the 'now' were not different but one and the same, there would not have been time, so too when its difference escapes our notice the interval does not seem to be time. If, then, the non-realization of the existence of time happens to us when we do not distinguish any change, but the soul seems to stay in one indivisible state, and when we perceive and distinguish we say time has elapsed, evidently time is not independent of movement and change. It is evident, then, that time is neither movement nor independent of movement."

But this argument is based on a confusion. He is using two words, μεταβολή and κίνησις for change, process or motion, and tends to use them sometimes in an inclusive sense in which the activity of conscious awareness constitutes a sort of change, and sometimes in a restricted sense in which only continuous physical processes count. In the former sense it is a truth, which even Barrow would concede [9], that in the absence of change we have no warrant for supposing any passage of time. If no physical change has occurred and no conscious mind has been aware of the passage of time, then it is an empty claim to assert that some interval of time has elapsed. But since he is going explicitly to allow that "καὶ γάρ ἐὰν ᾖ σκότος καί μηδὲν διὰ τοῦ σώματος πάσχωμεν, κίνησις δέ τις ἐν τῇ ψυχῇ ἐνῇ, εὐθὺς ἅμα δοκεῖ τις γεγονέναι καὶ χρόνος" [10], he cannot use κίνησις in the restricted sense of continuous physical process [11].

Although Aristotle's argument is confused, it raises the fundamental question 'How could one know, in the absence of change, that any time had elapsed?' We might reply that since we can know of things changing we can know of their not changing; and since sometimes, especially at night or during concentrated thought, our knowledge of the external world is limited, it is possible that every external thing we are conscious of is still, and that we are conscious of it as being still. This is the view of Plotinus and Henry More, and so, in part, of Barrow and Newton. "ἐν χρόνῳ γὰρ καὶ αὕτη [i.e. ἡ σφαῖρα] καὶ ἔστι, καὶ κινεῖται, κἄν στῇ, ἐκείνης [i.e. τῆς ψυχῆς] ἐνεργούσης, ὅση ἡ στάσις αὐτῆς, μετρήσομεν ...." ("The circle of the heavens itself exists and moves in time, and if it stands still we shall measure how long it stays stationary by the activity of soul ....") [12]. But the modern philosopher is loth

[9]   Though perhaps not Plotinus; see *Enneads*, III, 7, 9, ll. 79–84:

Εἰ μή τις τὴν γένεσιν αὐτοῦ παρὰ ψυχῆς λέγοι γίνεσθαι. 'Επεὶ διά γε τὸ μετρεῖν οὐδαμῶς ἀναγκαῖον εἶναι· ὑπάρχει γὰρ ὅσον ἐστί, κἄν μή τις μετρῇ. Τὸ δὲ τῷ μεγέθει χρησάμενον πρὸς τὸ μετρῆσαι τὴν ψυχὴν ἄν τις λέγοι·

which may be paraphrased: "Time originates from mind: not, however, because a mind is needed in order to measure it – it exists however much there is of it even if no one measures it. Rather, the mind *uses* magnitude to *measure* time."
[10]   Aristotle, *Physics*, IV, 11. 219a 4–6: "Even when it is dark, and we are not being affected through the body, if any process goes on in the mind, immediately we think that some time has elapsed too."
[11]   ibid. 219a 10–11.
[12]   *Enneads*, III, 7, 12, ll. 16–19. Compare Henry More's contention against Descartes (Second Letter, 5 March 1649, More to Descartes, in Henry More, *Collection of Severall Philosophical Writings* (London, 1662), p. 73; quoted by A. Koyré, *From the Closed World to the Infinite Universe* (Baltimore, 1957), p. 120, from Descartes, *Oevres*, ed. C. Adam and P. Tannery, (Paris, 1903), vol. V, p. 302): "*Nam si Deus hanc mundi universitatem annihilaret, et multo post aliam crearet de nihilo*, Intermundium illud, *seu absentia mundi*, suam haberet durationem quam tot dies, anni, vel secula mensurassent.*" ("For, if God annihilated this universe and then, after a certain time, created from nothing another one, this *intermundium* or this absence of the world would have its duration which would be measured by a certain number of days, years or centuries.") But More reifies (or, indeed, deifies) time too much, and in this last assertion he was wrong (see below § 56, p. 311), and needed to be put right by Barrow. See also St Maximus Confessor, *Ambiguorum Liber*, VI, 31, *P.G.* 91, 1164B14-C3; quoted below § 55, p. 303.

to accept the simple say-so of any individual as conclusive, and will demand evidence that our private experience was not illusory: and the only evidence he will accept is that of some change having occurred, contrary to our hypothesis. It is the same as with phenomenalism. To prove the passage of time in the absence of change is as difficult as to prove the existence of objects in the absence of their being observed. But the sceptic is being unreasonably hard to satisfy. In any particular case, the report of the elapse of time experienced without there being any change so far as the person was concerned may be mistaken. Barrow himself lays emphasis on 'more perfect', and allows, with Aristotle and Lucretius, that a human mind may be unable to perceive the passage of time in the absence of external change. But not everyone need be mistaken. On some occasions, in particular, we are conscious of everything being still, and later, when our experience is put in question, we discover that, unknown to us, some change had been going on – that the Pleiades had moved behind the spreading elm. This is enough to vindicate our original experience as having been veridical and to reject gratuitous imputations of unreliability. We can then affirm again that it is possible that every external thing we are conscious of is still, and that we are conscious of it as being still. It is only sometimes, to ward off the suggestion that our experience might have been illusory, that we need refer to some subsequent discovery of there having been a change going on unnoticed while everything for us was motionless. The state of perfect tranquillity is not only logically possible, but capable of being known by us individually, although epistemological difficulties prevent us proving to the sceptic the fact of our actually having had our calm hours. No contention can compel this conclusion, but only actual experience or the peaceful persuasion of the poets. We appeal to Sappho's silent solitude, and say we each can know that time steals by when all is still.

In fact, sceptical purposes apart, it is acknowledged. Even those philosophers who would banish time from the external world feel obliged to admit a phenomenal time as a concomitant of consciousness:

> The objective world simply *is*, it does not *happen*. Only to the gaze of my consciousness, crawling upward along the life line of my body, does a section of this world come to life as a fleeting image in space which continuously changes in time. [13]

Whatever else we think about changes in time, at least it must be conceded that sentient beings are under the impression that it passes. This is not to say

[13]   Hermann Weyl, *Philosophy of Mathematics and Natural Science* (2nd ed. New York, 1963), p. 116; criticized by Milič Čapek, *The Philosophical Impact of Contemporary Physics* (Princeton, 1961), p. 165. But see further Adolph Grünbaum, "Carnap's Views on the Foundation of Geometry", § IV, n. 102, in P. A. Schilp (ed.), *The Philosophy of Rudolph Carnap* (La Salle, 1964), p. 659; reprinted in J. J. C. Smart, *Problems of Space and Time* (New York, 1964), pp. 424 f., n. 35. See also O. Costa de Beauregard, "Relativity Theory: A Philosophy of Being", tr. David Park, in J. T. Fraser, *The Voices of Time* (London, 1968), pp. 429–30.

that we have special temporal experiences, comparable with the visual or olfactory sense data which the phenomenalists say we have. If we must talk of senses at all, we should – to use an old-fashioned term – ascribe temporal experiences to some inner sense rather than to any outer one. But it is better not to parcel out all experiences among the senses; and all we really want to say is that if a person is conscious at all, he is *eo ipso* conscious of the passage of time. Plotinus and St Augustine were right to maintain that time was a phenomenon of mind, and not merely a bare phenomenon but an extension, an experienced magnitude [14]. That time is an extension, that is to say, that with time it is proper to ask the question 'How much?', and answer it in terms of some amount of time, is to many people self-evident; but if further proof be wanted, it can be furnished, paradoxically, from consideration of the difficulties that arise from regarding it so. Although we are each conscious of the passage of time, we do not share a common experience of the measure of time. The passage of time is experienced by each individual as the sort of thing to which we can meaningfully ascribe a magnitude, but we have no direct awareness of the *measure* of time. We could not understand Juliet when she says

> I must hear from thee every day i' the hour,
> For in a minute there are many days: [15]

if her time were not to be measurable at all, nor again if the measure of time were known directly and minutes and days experienced as such, the former as one 1440th part of the latter. Only if time has the order-type of the continuum – the order-type (*Ordnungstypus*) θ in the language of the mathematicians [16] – can we be both sure that a measure can intelligibly be ascribed to it, and unsure what measure actually should be ascribed in some particular case. The question 'How long?' can always be asked, but the answer, given by the individual in terms of his own experienced awareness, characteristically differs from person to person and for the same person from time to time. Sometimes, time seems to stand still; at other times, the moments are so fleeting that it is impossible to experience them to the full. Paradoxically, it is times that at the time seemed to

---

[14]   Plotinus, *Enneads*, III, 7, 11, ll. 58–62:

Δεῖ δὲ οὐκ ἔξωθεν τῆς ψυχῆς λαμβάνειν τὸν χρόνον, ὥσπερ οὐδὲ τὸν αἰῶνα ἐκεῖ ἔξω τοῦ ὄντος, οὐδ' αὖ παρακολούθημα οὐδ' ὕστερον, ὥσπερ οὐδ' ἐκεῖ, ἀλλ' ἐνορώμενον καὶ ἐνόντα καὶ συνόντα, ὥσπερ κἀκεῖ ὁ αἰών.

("One must not take time to be external to the mind . . . nor merely the effect of it or consequent upon it, . . . but rather something appearing in the mind, inherent in it, and consubstantial in it.") See also *Enneads*, III, 7, 11, ll. 80–5, quoted on p. 12 above, and St Augustine, *Confessions*, bk XI, ch. XXVI, xxxiii: "*inde mihi visum est nihil esse aliud tempus quam distentionem: sed cuius rei, nescio, et mirum si non ipsius animi.*" ("Hence it seems to me that time is nothing other than extension; but extension of what I do not know, and wonder if it is not extension of mind itself.") And *Confessions*, bk IX, ch. XXVII, xxxvi: "*In te, anime meus, tempora mea metior.*" ("In the, my mind, I measure all my time.")
[15]   *Romeo and Juliet*, Act III, Scene V.
[16]   G. Cantor, *Math. Ann.* 46 (1895), p. 497; and, more fully, E. V. Huntington, *The Continuum* (2nd ed. Cambridge, Mass., 1917), ch. V, esp. § 54.

be dragging that in retrospect appear to have slipped by fastest, and times that sped by all too quickly at the time that subsequently seem to have been lengthy and full. There seems to be a general speeding up of the passage of time in the course of each person's life. Tomorrow to children never comes; old men speak of yesteryear as though it were the other day.

To talk of time going fast or slow is itself paradoxical. Many philosophers have noticed it, and have sought to forbid these locutions altogether. Nevertheless the urge to use them remains, because it stems from an ineliminable contrast between our subjective sense of time as experienced or remembered, and objective time as measured by calendars and clocks. We need to set up an objective, measured time because our individual senses of the passage of time do not agree: we can set up an objective, measured time because although we do not agree about the *extent* or *amount* of time that has passed, we do agree about the *order* in which events occur. An interview may seem an age to the man interviewed and only a few minutes to those giving the interview; but all will agree that the man interviewed came into the room and was invited to sit down and have a cigarette *before* he was told he could go and left the room.

Public or objective time is constructed on these two bases: each person is conscious of an *amount* of time passing, but does not naturally agree with others on how much time has passed; whereas we all do, normally and naturally, agree about the order in which temporal events occur. We are therefore able to select any suitable measurable process or succession of events, and agree between ourselves to make this our measure of public time. What standard we select is, within very wide limits, a matter of arbitrary choice. The ancient Romans used what we should regard as a variable hour – a twelfth part of daylight – which varied from season to season. The only two things that matter are that the process should be felt to be a process – something that occupies an amount of time – and that it should have a recognizable beginning and end (and perhaps some other definite stages in between), which everybody can recognize and use as *markers* in his own temporal consciousness. No one of these measures is absolute. There may be reasons for preferring one standard rather than another, but it remains always a matter of convention which standard we adopt. Whatever standard we do adopt, it will provide us with what we can regard as *isochronous intervals*; that is to say, we are to count the interval between two markers (e.g. two ticks of a clock or two successive noontides) *as* being equal. It is not to matter whether two isochronous intervals seem equal to any particular person: he is to count them as being really equal, irrespective of how they seem. He can do this, because the order of the markers marking the beginning and the end of isochronous intervals is the same for him as for anybody else. There can be no irresoluble doubt over whether two markers are successive markers, or over which came first and which came next. The order of the markers is intersubjective, and can, by convention, be made to lay off isochronous intervals. And so we set up objective, measurable time which we can all agree about and can all measure; and if it does not correspond to our own subjective experience of the passage of time, we can express the discrepancy by saying that time seems to be going fast, or passing slowly.

The discrepancy between "public" and "private" time and the consequent urge to talk paradoxically about time going fast or time going slow both arise from the fact that while we each experience time as an amount – a magnitude – all that we can always agree about is an order. Subjective time, as we have said has order-type θ – the order-type of the real numbers: but all that intersubjective time appears to have is order-type η [17] – the order-type of the rational numbers (see further below §§ 5, 6, esp. p. 30). We have no naturally given intersubjective experience of the amount of time elapsing, but because of our subjective experience of it, we are sure that intervals of time do have magnitudes, and therefore insist that the question 'How much?' shall be askable of public time as well as of private. If our subjective experience of time were less firmly of order-type θ, we should not be so anxious to wish order-type θ on public time too; if our experience of the passage of time were less subjective, we should have less difficulty and need less artifice in making out public time to be a magnitude. But as it is, our experience of time is enough to make us want to be able always to regard it as a magnitude, but not enough to enable us to do so naturally and without danger of paradox.

[17]   See, again, E. V. Huntington, op. cit., ch. IV, esp. §§ 144, 151.

§ 3
Instants and intervals

The account just given of public time depends on a distinction between the *instants* that mark the beginning and end of isochronous intervals and the *intervals* that lie between these instants. The distinction between instants and intervals is one that is difficult to draw clearly. It is dangerously easy to suppose that instants must be extremely short intervals, and that intervals must be composed of a large number of instants. It is this supposition that lies behind the paradoxes of Zeno [1] and, more excusably, the difficulties St Augustine gets himself into [2]. Ideally, instants are like points; they have no extension or magnitude. We can ask where a point is, and whether it lies on a certain line or not, and whether it lies within a certain figure or not; but we cannot ask how large a point is – not when we are considering ideal points. Intervals, on the other hand, can have magnitude, just as segments of a line can. We can sensibly ask the question 'How much?' of an interval, even if we cannot always give an answer.

The analogy between time and space, although useful here, is dangerous. Fortunately we can explain the ideal concepts of instant and interval not by

[1]   Aristotle, *Physics*, VI, 9; see also VI, 2, 233a 13 ff., and VIII, 8, 263a 4 ff.
[2]   St Augustine, *Confessions*, bk XI, chs. XII–XXIII, xiv–xxix.

analogy with point and segment, but in terms of the *pure* mathematics of the real-number continuum. An instant corresponds to a real number; a temporal interval corresponds to an interval of real numbers. We do not think of real numbers as very small intervals of the real-number continuum, and this will help us not to think of ideal instants as very small intervals of time. We accept that in a finite interval of the real-number continuum there are an infinite number of real numbers, and this does not lead to a contradiction because each real number by itself is not an interval at all; and if it is further asked how even an infinite number of magnitude-less entities can constitute an interval that does have magnitude, we can do something to mitigate the difficulty by adducing Cantor's proof that the number of real numbers in an interval of the continuum is a *larger* infinity than the denumerable infinity of our ordinary thinking [3]. In exactly the same way we may now accept that in a finite interval of time there are an infinite number of instants, and this does not lead to a contradiction because each instant by itself is not an interval at all, and does not take any time to elapse [4]. Zeno's arguments, about Achilles and the Tortoise and about the Arrow, lose all force. In so far as they are difficulties at all they are difficulties about the concept of the continuum, and not about time as such. We may well concede that the concept of magnitude, quantity or amount does indeed involve us in all the difficulties that the Greeks detected: but, thanks to the work of Cantor and Dedekind, these difficulties can be resolved.

Cantor and Dedekind adopted different approaches to the definition of a real number. Cantor defined a real number not as an interval but as an infinite series of "nested intervals" of rational numbers: Dedekind defined a real number as a "cut" or division (which could be regarded as a special sort of infinite interval) of rational numbers. Both approaches have been foreshadowed in traditional puzzles about the nature of time, which are familiar to us in St Augustine's argument of the ever-shrinking present and two of his doctrines of the ever-present present. These puzzles can be resolved by use of the distinction between instants and intervals, and provide in effect a proof, one like Cantor's, another like Dedekind's, that time is a continuum of order-type θ, and not, as many philosophers, especially Hume and the empiricists, have supposed, a succession of discrete moments of order-type ω (like the natural numbers, 1, 2, 3, ...: i.e. with a beginning but no end) or $\omega^* + \omega$ (like the integers ... −3, −2, −1, 0, 1, 2, 3, ...: i.e. with no beginning and no end) [5].

St Augustine's argument of the ever-shrinking present will be discussed first, as the one to which the distinction between instants and intervals is most immediately relevant. Next, empiricist and then other objections to infinite

[3]   For a proof, see e.g. E. Kamke, *Theory of Sets*, tr. F. Bagemihl (New York, 1950), ch. I, § 4, pp. 9–10; or S. C. Kleene, *Introduction to Metamathematics* (Amsterdam, 1952), ch. I, § 2, pp. 6–7; or Geoffrey Hunter, *Metalogic* (London, 1971), § 11, pp. 22–4.
[4]   Aristotle had realized this, although he did not make full use of it. See *Physics*, IV, 11, 220a 17–20.
[5]   E. V. Huntington, *The Continuum* (2nd ed. Cambridge, Mass., 1917), ch. III.

divisibility of time will be discussed and met. Denseness established, Dedekind's argument will be deployed to give a formal proof that we need to ascribe to public time, as well as to private time, the order-type θ of the continuum of real numbers. We can then use the distinction between instants and intervals to give a concise account of the construction of objective time and the problem of measuring it. Each person experiences his own, subjective time as an interval, a continuum of order-type θ, about which it makes sense to ask 'How much?'. The topological properties of different people's temporal experience are invariant, but not the metrical properties. We agree about instants but not about intervals. We therefore establish a metric by adopting a suitable continuous process as a standard; this we can regard as a continuum articulated into a number of intervals, whose topology will be the same for all, and which can therefore be taken as defining for each person equal temporal intervals in his own experience.

§ 4
The ever-shrinking present

St Augustine allowed himself to be confused about the present. He may be forgiven. He was impelled to maintain the present both to be on a par with the past and future, and therefore to be an interval, and to be in a special category of its own, and to be an instant, the unique instant dividing the future from the past. Both views may be maintained, but not both at once, or inconsistency is bound to result. Nevertheless, St Augustine may be forgiven. There are strong linguistic pressures, even stronger in Latin and Greek than in English, in both directions, and quite apart from the problem of whether we should regard the present as an interval or as an instant, there are difficulties about the meaning of the words 'past, 'present' and 'future'. They depend for their force on their context of utterance and thus have built into them a covert egocentricity of meaning. Crudely:

The future			instants or an interval after		
The	}	refers to	the instant itself or an	}	the instant at
present			interval including		which I am
The past			instants or an interval before		speaking

These words, along with the pronouns 'I', 'now', 'here', 'this', 'that', 'then', 'there', are "token-reflexive" as Reichenbach calls them or "egocentric particulars" according to Russell's terminology [1]. Discourse about the Present, the Future and the Past, and arguments about whether the Future will resemble the Past, are dangerous, because, in spite of their capital letters, the Present, the Future and the Past, may not be well-formed expressions. Some philosophers have even argued that the traditional sceptical thesis about induction cannot be maintained because it cannot be stated in non-egocentric language [2]. That goes too far. But it is important always to beware of the covert egocentricity of these words (see further below, § 30, pp. 144–5, § 44, p. 216, § 52, § 53, § 55, pp. 304–6).

Many of these words, in particular the words 'now', 'then' and 'present', depend on their context in a further respect. Consider first the spatial words 'here' and 'there'. They indicate only roughly a *region round* a given point or in a given direction. What the region is depends very much on the context of utterance. In the same way the adverb 'now' and the adjective 'present' are used to refer to a somewhat indefinite interval around the instant of utterance. How large this interval is depends on the context, and the way is open for St Augustine's argument:

> *Videamus ergo, anima humana, utrum praesens tempus possit esse longum: datum enim tibi est sentire moras atque metiri. quid respondebis mihi? an centum anni praesentes longum tempus est? vide prius, utrum possint praesentes esse centum anni. si enim primus eorum annus agitur, ipse praesens est nonaginta vero et novem futuri sunt, et ideo nondum sunt: si autem secundus annus agitur, iam unus est praeteritus, alter praesens, ceteri futuri. atque ita mediorum quemlibet centenarii huius numeri annum praesentem posuerimus: ante illum praeteriti erunt, post illum futuri. quocirca centum anni praesentes esse non poterunt. vide saltem, utrum qui agitur unus ipse sit praesens. et eius enim si primus agitur mensis, futuri sunt ceteri, si secundus, iam et primus praeteriit et reliqui nondum sunt. ergo nec annus, qui agitur, totus est praesens, et si non totus est praesens, non annus est praesens. duodecim enim menses annus est, quorum quilibet unus mensis, qui agitur, ipse praesens est, ceteri aut praeteriti aut futuri. quamquam neque mensis, qui agitur, praesens est, sed unus dies: si primus, futuris ceteris, si novissimus, praeteritis ceteris, si mediorum quilibet, inter praeteritos et futuros.*
>
> *Ecce praesens tempus, quod solum inveniebamus longum appellandum, vix ad unius diei spatium contractum est. sed discutiamus etiam ipsum, quia nec unus dies totus est praesens. nocturnis enim et diurnis horis omnibus viginti quattuor expletur, quarum prima ceteras futuras habet, novissima praeteritas, aliqua vero interiectarum ante se praeteritas, post se futuras. et ipsa una hora*

[1]   Hans Reichenbach, *Elements of Symbolic Logic* (New York, 1951), § 50, pp. 284–7; Bertrand Russell, *Enquiry into Meaning and Truth* (London, 1940), ch. VII, p. 128, or *Human Knowledge: Its Scope and Limits* (London, 1948), pt. II, ch. IV, p. 100.
[2]   A. J. Ayer, *Philosophical Essays* (London, 1963), p. 188; *Problem of Knowledge* (London 1956), p. 160. S. N. Hampshire, "The Analogy of Feeling", *Mind*, LXI (1952), pp. 7–12.

fugitivis particulis agitur: quidquid eius avolavit, praeteritum est, quidquid ei restat, futurum. si quid intellegitur temporis, quod in nullas iam vel minutissimas momentorum partes dividi possit, id solum est, quod praesens dicatur; quod tamen ita raptim a futuro in praeteritum transvolat, ut nulla morula extendatur. nam si extenditur, dividitur in praeteritum et futurum: praesens autem nullum habet spatium. ubi est ergo tempus, quod longum dicamus? an futurum? non quidem dicimus: longum est, quia nondum est quod longum sit, sed dicimus: longum erit. quando igitur erit? si enim et tunc adhuc futurum erit, non erit longum, quia quid sit longum nondum erit: si autem tunc erit longum, cum ex futuro quod nondum est esse iam coeperit et praesens factum erit, ut possit esse quod longum sit, iam superioribus vocibus clamat praesens tempus longum se esse non posse.

Let us see therefore, O thou soul of man, whether the present time may be long. For to thee it is given to be sensible of the distances of time, and to measure them. What now wilt thou answer me? Are an hundred years in present a long time? See first, whether an hundred years can be present. For if the first of these years is now current that one is present indeed, but the other ninety and nine are to come, and therefore are not yet: but if the second year is now current, then is one past already, another in present being, and all the rest to come. And if we suppose any middle year of this hundred to be now present; all before it will be past, all after it to come. Wherefore an hundred years cannot be present. See again whether that one which is now current be now present. For even of that, if the first month be now current, then are all the rest to come: if the second, then is the first past, and the rest not yet come on. Therefore, neither is the now current year all present together: and if it be not all present, then is not the year present. For twelve months are a year; of which that one now current is present, all the rest either past, or to come. Yet neither is that month now current, present; but one day of it only: if the first, the rest are to come; if the last, the rest are past; if any of the middle, that is that between the past and the future.

See how the present time (which alone we found meet to be called long) is now abridged to the length scarce of one day. But let us examine that also; because not so much as one day is wholly present. For four and twenty hours of day and night do fully make it up: of which, the first hath the rest to come; the last hath them past; and any of the middle ones hath those before it already past, those behind it yet to come; yea, that one hour is wasted out in still vanishing minutes. How much so ever is flown away, is past; whatsoever remains, is to come. If any instant of time be conceived, which cannot be divided either into none, or at most into the smallest particles of moments; that is the only it, which may be called present; which little yet flies with such full speed from the future to the past, as that it is not lengthened out with the very least stay. For lengthened out if it be, then is it divided into the past and the future. As for the present, it takes not up any space. Where then is the time which we may call long? Is it to come? Surely we do not say, that that is long: because that of it is not yet come which may be long: but we say, it will be long. When therefore will it be? For if even then it is yet to come, it shall not be long, because there will be not yet that which may be long: but if it shall then be long, when from the future (which it is not yet) it shall begin now to be, and shall be made present, that so there may now be that which may be long, then does the present time cry out in the words above rehearsed, that it cannot be long. [3]

Augustine's argument depends essentially on the fact that the word 'present' can mean 'this century' or 'this year' or 'this month' or 'this day' or 'this minute' or even 'this second'. It can mean any of these, and we may be inclined to hold, as St Augustine did, that the latter uses of the word are more stringent, and somehow better, than the former. And then we may be tempted to extrapolate and maintain that the real meaning of 'present' is the smallest possible interval around the given instant. But there is no smallest possible interval; there never are, with intervals. And so we may be forced to the conclusion that there is no time that the word 'present' can really refer to, and so that time is unreal.

The solution to this problem of the ever-shrinking present is secured if we maintain a sharp distinction between instants and intervals, and distinguish uses of the word 'present' in which it is being used in order to refer to an instant, and uses in which it is being used in order to refer to intervals. If I am talking of instants, I can among other instants talk of the present instant: I am more likely, however, to use the word 'now' to refer to an interval; but then I must not be ashamed of the interval's being an interval, and not try and make it very small, small enough to count as an instant instead. To prevent confusion, philosophers should avoid the word 'now' [4] and use the phrase 'the present',

[3]   St Augustine, *Confessions*, bk XI, ch. XV, xvii–xx (English translation from the Loeb edition). The argument is given also by Chrysippus *apud* Stobaeus, *Eclogae*, I, p. 106, 5W (*Arii Didymi*, fr. 26, p. 164, ll. 25–30, ed. Diehls); reprinted in J. von Arnim, *Stoicorum Veterum Fragmenta* (Leipzig, 1903), II, 509; and by Sextus Empiricus, *Adversus Mathematicos*, X (i.e. Πρὸς φυσικούς, B), 182–4; and alluded to in part by Aristotle, *Physics*, IV, 10, 217b 33–5.
[4]   This was more important in antiquity, where Latin and Greek each have two words – *nunc, νῦν*, and *jam, ἤδη* – which just fail to draw the crucial distinction. There is a tendency (especially in Attic Greek) to use the former to refer to the present instant, and the latter the present interval. Aristotle, *Physics*, IV, 13, 222b 7–11, attempts to draw this distinction:
τὸ δ' ἤδη τὸ ἐγγύς ἐστι τοῦ παρόντος νῦν ἀτόμου μέρος τοῦ μέλλοντος χρόνου (πότε βαδίζεις; ἤδη, ὅτι ἐγγὺς ὁ χρόνος ἐν ᾧ μέλλει), καὶ τοῦ παρεληλυθότος χρόνου τὸ μὴ πόρρω τοῦ νῦν (πότε βαδίζεις; ἤδη βεβάδικα).
The word ἤδη refers (i) to the part of the future that is near to the instantaneous present νῦν – e.g. "When are you walking?" "ἤδη" – because the time in which he will be walking is close at hand; and (ii) to the part of the past, which is not far from the νῦν – e.g. "When do you walk?" "I have been walking ἤδη." The Greek cannot be rendered accurately into idiomatic English. 'Presently' has lost its reference to the recent past in modern English. 'Just' is more idiomatic – 'When are you coming?' 'I am just coming/I have just come' – but

and remember that it is an adjective and ask themselves what the noun is that is being understood. If it is the present instant, well and good; if it is the present century, well and good; if it is the present month, well and good. All these are permissible and intelligible locutions, and none is better or more exact than any other. The only thing to remember is that whereas the present month is about 1:1200 part of the present century, the present instant is no part whatever of the present month. Or, to put it another way, we can if we like have the present *pari passu* with the past and the future, referring, like them, to an interval: or we can have the present being unlike the past and the future, being in fact the boundary between the past and the future; but not both.

The trouble is, we want to have both. An instantaneous 'now' by itself is not enough; for one thing, we can never point it out with absolute accuracy, and there are other conceptual and psychological pressures in favour of a "specious" present that must be some sort of interval: but if an account of the present in terms of intervals alone is offered, we find we need an instantaneous 'now'. The "specious present" is a conceptual necessity in so far as we consider time from the standpoint of the agent. Deeds are not discrete. We cannot analyse activities into series of isolated atomic acts. Rather, the correct description of what a person is doing at any definite instant will depend on what he has been doing and is going to do. The concept of action thus imports a temporal indefiniteness into our language and thought, which is naturally and properly agent-oriented. Even when we are not acting, but only perceiving, we experience the present more as an interval than an instant. It is difficult to be clear about the phenomenon of the specious present, partly because it is inherently indeterminate. As Whitehead says, "The temporal breadths of the immediate durations of sense-awareness are very indeterminate and dependent on the individual percipient. ... What we perceive as the present is the vivid fringe of memory tinged with anticipation" [5]; but fringes, if uncertain intervals, are certainly not instants. And therefore an instantaneous 'now' is empirically inadequate. It remains, however, theoretically indispensable. For behind St Augustine's dialectic lies the mathematical fact that an infinite set of nested present intervals defines a present instant. Indeed, Chrysippus almost anticipated Cantor, saying that the present was only "loosely defined" and describing "the Now as a limiting point between nested intervals bounded on

---

carries with it overtones of discreteness we are anxious to avoid. Nor is the distinction universally observed in Greek: in *Iliad*, 3,439, νῦν Μενέλαος ἐνίκησεν, *Iliad*, 5,279, νῦν αὖτ' ἐγχείη πειρήσομαι, and elsewhere, νῦν and *nunc* are used to refer to intervals as well as instants, as Aristotle himself notes earlier in the same chapter, 222a 20–2: "τὸ μὲν οὖν οὕτω λέγεται τῶν νῦν, ἄλλο δ' ὅταν ὁ χρόνος ὁ τούτου ἐγγύς ᾖ. ἥξει νῦν, ὅτι τήμερον ἥξει· ἥκει νῦν, ὅτι ἦλθε τήμερον." And as his elucidation shows, ἤδη tends to be used for the present interval excluding the present instant; that is, the so-called immediate future and immediate past. Or in Latin, as Alice sadly learnt, "jam tomorrow, jam yesterday, but never jam today". Thus *jam* and ἤδη are not available for referring to the present interval, and *nunc* and νῦν could never be entirely reserved for the instantaneous present.
[5]   A. N. Whitehead, *The Concept of Nature* (Cambridge, 1920), pp. 72 f.

the one side by the past and on the other by the future" [6]. We therefore cannot follow Findlay, and put some definitional stop to Augustinian perplexities [7]. Once we allow that the present is an interval at all, then whatever interval is considered, it can be considered as made up of smaller intervals, of which only one can be regarded as the present interval, and all the others either past or future. This process of precisification can be continued indefinitely, and has as its limit not an interval at all, but a durationless instant. If we attempt to find a sticking point anywhere short of the limit, we shall be dislodged by the argument St Augustine originally deployed: if we go along to the limit, we are likely to be perplexed by the somewhat counter-intuitive fact that a denumerably infinite intersection of intervals may yield an instant, whereas a denumerably infinite union of instants cannot constitute an interval. This is a piece of pure mathematics. It does not depend on any property of time. But it does mean that if we are going to think consistently about time, we need to take great care to distinguish instants from intervals, and in particular to distinguish the present instant from the various intervals that it is contextually appropriate to call present.

[6]   S. Sambursky, "The Concept of Time in Late Neoplatonism", *The Israel Academy of Sciences and Humanities, Proceedings*, vol. 11, No. 8, pt 2, based on Stobaeus, *Eclogae*, I, 106 (J. von Arnim, *Stoicorum Veterum Fragmenta* (Leipzig, 1903), II, 509):
'Ἐμφανέστατα δὲ τοῦτο λέγει, ὅτι οὐδεὶς ὅλως ἐνίσταται χρόνος. Ἐπεὶ γὰρ εἰς ἄπειρον ἡ τομὴ τῶν συνεχόντων ἐστί, κατὰ τὴν διαίρεσιν ταύτην καὶ πᾶς χρόνος εἰς ἀπειρονέχει τὴν τομήν· ὥστε μηθένα κατ' ἀπαρτισμὸν ἐνεστάναι χρόνον, ἀλλὰ κατὰ πλάτος λέγεσθαι.
I am not sure how far κατὰ πλάτος will bear a Cantorian interpretation, but there seems to be some support from Proclus in *Platonis Timaeum*, p. 271 (J. von Arnim, op. cit., II, 521), who says that the Stoics accounted time ἀμενηνὸν καὶ ἔγγιστα τοῦ μὴ ὄντος; the superlative gives some suggestion of a limit of nested intervals.
[7]   J. N. Findlay, "Time: A Treatment of Some Puzzles", *The Australasian Journal of Psychology and Philosophy*, XIX (1941), pp. 225–7; reprinted in A. G. N. Flew, *Logic and Language*, series I (Oxford, 1951), pp. 45–7; and in J. J. C. Smart, *Problems of Time and Space* (New York, 1964), pp. 346–8.

§ 5
Concepts and experience

The theory of real numbers gives us a consistent and complete framework for handling the concept of quantity. A quantity is to be represented as lying in a continuous series of order-type θ, the order-type of the real numbers. But although this provides by far the most satisfactory conceptual framework, our practical applications are much cruder than the theory of real numbers would suggest. We may calculate that the diagonal of an isosceles right-angled triangle ought to be $\sqrt{2}$ units long, but we never measure it as being exactly that. Indeed, we never measure any continuous quantity as being *exactly* anything; it is always an approximation to within a greater or lesser degree of accuracy. The actual points we draw, as opposed to the ideal points we talk about, are never completely without magnitude; drawn lines have some breadth; the finest balances yet constructed will still represent as equal two weights which differ only by a sufficiently small amount; and we cannot indicate an instant with absolute precision, nor can we say which of two came first if they came sufficiently close together.

Empiricists, therefore, take exception to the application of the theory of real numbers to the world of experience, and deny that time can be infinitely divisible or that there can be mathematical instants of time as we have described them. Men, being only finite, cannot divide anything infinitely many times. We cannot really say that between any two dates there must always be a third, because our temporal discrimination is imperfect, and we cannot tell, with events that are nearly simultaneous, which came when – we have to rely on cameras, rather than our own judgement, to tell us in a photo-finish which horse came first; and it is quite wrong therefore to ascribe to our awareness of time any order-type that presupposes our being able to draw finer distinctions than in fact we can. We must not precisify our concept of time beyond the blur given to us in ordinary experience. Thus Hume:

'Tis universally allowed that the capacity of the mind is limited, and can never attain a full and adequate conception of infinity: And tho' it were not allow'd, it wou'd be sufficiently evident from the plainest observation and experience. 'Tis also obvious, that whatever is capable of being divided *in infinitum*, must consist of an infinite number of parts, and that 'tis impossible to set any bounds to the number of parts, without setting bounds at the same time to the division. It requires scarce any induction to conclude from hence, that the *idea*, which we form of any finite quality, is not infinitely divisible but that by proper distinctions and separations we may run up this idea to inferior ones, which will be perfectly simple and indivisible. [1]

It is a problem, not only about time, but about every sort of measurement, and the theory of knowledge generally. None of our measurements of continuous quantities are absolutely accurate. Instead of being able to compare two quantities and say that one is greater than the other, or else they are both equal, we can in fact say only that one is *noticeably* greater than the other, or else that neither is noticeably greater than the other – that is, they are *approximately* equal. We have to work with the concept 'not noticeably greater than' instead of the more exact and mathematically more convenient one 'not greater than'; among other disadvantages, the valid argument that two things that are equal to a third are equal to each other is no longer valid when we substitute 'approximately equal' for 'equal'.

By careful use of the concept 'not noticeably greater than' we could build up a complete theory of measurable quantities; or we can give application rules for the application of an existing theory with its ideal entities and ideal quantities within certain margins or error. The disadvantages of doing the former are, first, that such a theory will be very clumsy and, secondly, that the 'noticeably' of 'not noticeably greater than' is indefinite: it varies from occasion to occasion; in particular, it changes with each new refinement of measuring technique. Once cameras have been invented, we use them to determine the order in which horses arrive in a photo-finish. We have no hesitation in accepting the results yielded by the camera, any more than with those yielded by the magnifying

[1] David Hume, *Treatise on Human Nature*, ed. L. A. Selby-Bigge (Oxford, 1888), bk I, pt II, § I, pp. 26–7.

glass or Vernier gauge. Our inability to discriminate we ascribe entirely to ourselves and our own imperfect organs, and not to any inherent indivisibility of what we are failing to divide. The very examples cited by the empiricists to show the existence of a *minimum sensibile* are incoherent, and in saying how small the smallest *per*ceivable percept is, show that a smaller one still is *con*ceivable, and might conceivably, by some man, be perceived. We are often prepared to say that there must have been a difference, though it was too small to be noticed, rather than abandon some other principles; for example, if we weighed one mass against another, and found them equal, and the second against a third and so on, and finally found that the last mass did not balance against the first, we should invoke the possibility of there having been differences too small to be noticed rather than abandon the principle of transitivity. This shows that the concept of being not noticeably greater than is too indefinite and soft to bear the weight of a theory, and that we have to construct our theories in terms of ideal entities and quantities, and only when we come to apply them in a particular case to consider the necessarily approximate character of our measurements. This distinction, between *concepts* and the *criteria for applying them in particular cases* is of fundamental importance throughout philosophy. We have already noted (§ 2, p. 9) how failure to draw this distinction leads to absurd consequences when we consider the concept of probability or the concept of pain, and we shall have many occasions to revert to this distinction in our elucidation of the concepts of space and of time (§ 11, pp. 67–8, § 25, pp. 119–20, § 30). When we raise the question of the infinite divisibility of time, what we are concerned with is the concept of time; not how we can, as a matter of fact, discriminate small differences of time in our actual experience, but how we think of time. Our experience of time is, indeed, blurred, as the empiricists maintain: but our concept of time is of something infinitely divisible. For, however small an interval of time we think of, we should not regard it as an interval unless it had a beginning and an end, in which case we can also think of an instant after the beginning and before the end. Which is to say that we *think* of time as infinitely divisible.

§ 6
Denseness and continuity

It is now generally taken for granted that public time is both infinitely divisible, or "dense" as the mathematicians term it, and continuous; that is, not only can we always consider any interval as made up of smaller ones, but we are entitled to apply even irrational numbers to the measurement of time. If we discover that a freely falling body takes 2 seconds to fall 64·4 feet, we say that it would have taken $\sqrt{2}$ seconds to fall 32·2 feet. We are right to do so, but not self-evidently right. Quite apart from Zeno's difficulties with infinity, and Hume's empiricist objections, the notion of denseness and continuity is incompatible with the notion of *nextness*, and hence, on one of its interpretations, of *succession*. If time is dense, there cannot be a next instant after this one; because for any instant after this one there is, by the definition of denseness, *another* instant between it and this one; so that it could not have been the next one. But we often talk as if there were such contiguous instants "The next moment he seized the controls ...". More especially, if we regard time as a succession, and interpret successor to mean 'next after', we seem to be forced to the conclusion that time cannot be infinitely divisible, and must instead be made up of indivisible moments. Hume uses such an argument to reinforce his empiricist argument against infinite divisibility:

'Tis a property inseparable from time, and which in a manner constitutes its essence, that each of its parts succeeds another, and that none of them, however contiguous, can be coexistent. For the same reason, that the year 1737 cannot concur with the present year 1738, every moment must be distinct from, and posterior or antecedent to another. 'Tis certain then, that time, as it exists, must be composed of indivisible moments. For if in time we could never arrive at an end of a division, and if each moment as it succeeds another, were not perfectly single and indivisible, there would be an infinite number of coexistent moments, or parts of time; which I believe will be allow'd to be an arrant contradiction. [1]

The error lies in the word 'another'. It can mean 'any other', but it suggests 'just one other'. Every moment (in our sense of 'instant') is indeed distinct from, and posterior or antecedent to, any other moment, and each moment as it comes after other moments is perfectly single and indivisible. But since there is not just one moment that it succeeds (in the strong sense of coming next after), the inference that there would have to be an infinite number of co-existent moments does not hold, and the contradiction is avoided.

The order-types ω, η and θ, the order-types of the natural numbers, of the rational numbers and of the real numbers respectively, are all three *complete* orderings: that is, given any two numbers, either they are the same, or one comes after the other. There is no possibility, in those cases where any of these orderings apply, of there being two distinct members that are not related to each other by the ordering relation. There are other, the so-called "partial", orderings where comparisons cannot always be made. An economist may reasonably regard me as better off than an Indian peasant, and as worse off than the Duke of Bedford. But whether I am better or worse off than an American professor, a British civil servant, or even a fellow of another college, are questions to which there are no answers. The relation of being better off than is asymmetrical (if I am better off than you, then you are *not* better off than me), and transitive (if the Duke of Bedford is better off than me, and I am better off than an Indian peasant, then the Duke of Bedford is better off than an Indian peasant), and therefore is an ordering relation; but it gives only an incomplete ordering, since comparisons are not always possible. By contrast, temporal comparisons always are possible [2]; of any two, distinct, instants one must be after the other or vice versa, and this is to say that the relation 'after' gives a *complete* ordering of instants. Hume was right to think that our temporal ordering was complete, wrong to assume that it must therefore be like that of the integers or natural numbers.

These arguments against the infinite divisibility of time are invalid; but that is not enough to show that time is infinitely divisible. The neatest argument in

[1]   David Hume, *Treatise on Human Nature*, ed. L. A. Selby-Bigge (Oxford, 1888), bk I, pt II, § II, p. 31.
[2]   Provided they are at the same place. See below § 7, pp. 36 ff., and §§ 44, 46.

favour is Aristotle's [3]. Aristotle argues that provided we have two velocities (or, more generally, two rates of change of any, not necessarily spatial, magnitude), then if at the lower velocity it takes a certain interval of time to cover a given distance (or accomplish a given change), then at the higher velocity the *same* distance (or accomplishing the same change) will be covered in a lesser period of time; but in this lesser period of time, at the lower velocity a lesser distance will be covered, which lesser distance will be covered in even less time at the higher velocity. We can go on alternating thus, always dividing further any given interval. So time is infinitely divisible.

Aristotle's argument may be attacked in two ways, and the development of relativity theory and quantum mechanics lends some support to each of them. It may be questioned whether there really are two possible velocities. At first sight it would appear that there obviously were, and that we could prove it by an adaptation of one of Zeno's arguments. If we allow that we can have velocities in opposite directions, and consider the relative velocities of two bodies going with a velocity with respect to a stationary body, but in opposite directions, we should suppose that their relative velocity with respect to each other must have an absolute magnitude greater than that of their velocities with respect to the stationary object [4]. But the argument does not hold where the velocity of light is concerned. If two photons leave a source, each with velocity c but in opposite directions, then the velocity of one photon with respect to the other is still only c (see below § 42). The velocity of light is unique. And one Russian scientist has attempted to build up a theory in which time and space are not infinitely divisible but discrete, and the fundamental entities can move only with the velocity of light [5].

More plausibly, arguments from quantum mechanics can be adduced for there being an ultimately granular structure of time and space, with quanta of duration and distance, as well as of action. We can always attribute to the smoothing-out effect of great numbers the fact that movement, and change generally, appears to be continuous rather than jerky. Cine films are composed of large numbers of separate stills, but give the impression of continuous motion; and time too might be composed of discrete minimum moments, with non-infinitesimal changes occurring or not occurring from one to the next [6]. But quantum mechanics has not established this. What it has shown is that we cannot discriminate between instants separated by very small intervals, if the energy of the system is not to be implausibly imprecise. If the energy is to be reasonably definite – and a system in which it was not would scarcely count as a physical system at all – the interval over which it is measured cannot be

[3] *Physics*, VI, 2, 232a 23–233b 33.
[4] Quoted by Aristotle, *Physics*, VI, 9, 239b 33–240a 18.
[5] А. Н. Вяльцев, Дискретное Пространство-Время (Moscow, 1965).
[6] For a fuller discussion, see G. J. Whitrow, *The Natural Philosophy of Time* (Edinburgh, 1961), pp. 136–7.

made vanishingly small. But that is not to say that time is granular. If time were granular, moments or "epochs" or "chronons" would succeed one another the same for all systems: every system would have to be in step with every other one, and the Heisenberg uncertainty principle would cease to specify an ineliminable uncertainty of timing for a given system with its energy determined within given limits, and instead would specify for every system, irrespective of the extent to which its energy was determined, the temporal magnitude of their successive moments. It is not a natural interpretation of Heisenberg's principle. Even if we regard it as not only laying down limits on the accuracy of measurements we can make, but as ascribing some fundamental inexactness to the way things are, it does not show time to be granular, but only that it is blurred to a varying extent under varying conditions.

Although the present state of quantum mechanics does not warrant it, we might still be led to conclude that time was quantized [7]. Indeed, some physicists have suggested that each second is made up of $2 \cdot 2 \times 10^{23}$ chronons [8]. It could be so (but see further below, § 15, p. 85). It would be an exciting discovery in physics if it were. But it would be a discovery about the physics of time, not the concept of time. It would be a contingent truth that events could not occur at intervals of less than $4 \cdot 5 \times 10^{-24}$ seconds, but it would not show that time was composed of a discrete succession of instants or space of a discrete lattice of fundamental points: for the minimum intervals would be intervals still, and we could still ask how *long* a minimum interval lasted. We may compare time with mass, where it is a contingent and not a conceptual truth that mass is not infinitely divisible, and we never come across particles with a mass less than, say, that of the electron. We still can think of smaller masses, and may one day come across particles actually possessing them. With time (and distance) it seems far more certain that these are given us *as* quantities, not as successions, and that with these the right question is 'How much?' not 'How many?' As Kant put it: "Space and Time are *quanta continua*, because no part of them *can be given* [my italics] save as enclosed between limits (points or instants), and therefore only in such fashion that this part is itself a space or a time." [9] We might, conceivably, come to think differently, if the existence of chronons or epochs was well established by science. Our concepts are not immune to revision; and in the case of time, we are already prepared, in some locutions, to speak of it as though it were discrete. But to do so consistently would require a fairly radical revision of the concept. We

[7]   A. N. Whitehead, *The Concept of Nature* (Cambridge, 1919), p. 162.
[8]   J. J. Thomson, "The Intermittance of Electric Force", *Proceedings of the Royal Society of Edinburgh*, XLVI (1925–6), p. 90. E. Borel, *L'Espace et le temps* (Paris, 1923), pp. 124–7. R. Lévi, "La théorie de l'action universelle et discontinue", *Journal de Physique et de Radium*, VIII (1927), pp. 182 f. J. A. Wheeler and R. P. Feynmann, *Reviews of Modern Physics*, XVII (1945), pp. 157–81. Sir Edmund Whittaker, *From Euclid to Eddington* (2nd ed. New York, 1958), p. 41. Milič Čapek, *The Philosophical Impact of Contemporary Physics* (Princeton, 1961), pp. 230–8. G. J. Whitrow, *The Natural Philosophy of Time* (Edinburgh, 1961), pp. 153–7.
[9]   *Critique of Pure Reason*, tr. Norman Kemp Smith (London, 1929), A169/B211, p. 204.

should have to unthink at least as far back as Aristotle; as we shall see, other fundamental concepts presuppose the infinite divisibility of time (see below §26, p. 123, §29, pp. 136–7). The concept *we* have is of an infinitely divisible, or, as the mathematicians term it, dense, time. None of the arguments against infinite divisibility hold water; nor is there an alternative concept of discrete time readily available. These two considerations constitute for the present a decisive, although not necessarily for all time conclusive, argument for maintaining that public time *is* dense, with at least the order-type η, the order-type of the rational numbers.

In fact we want more. We want public time, like private time, to have the order-type of the continuum. For, surprisingly to the non-mathematician, even a dense ordering has gaps. It was the great intellectual scandal of the Greeks that there was no rational number whose square was exactly 2, although it follows from Pythagoras' theorem that such a magnitude must exist. For if there were such a rational number, it could be expressed as a ratio p/q of two natural numbers, p and q, having no factor common to them both. These two natural numbers would have to be such that $p^2 = 2q^2$, from which it would follow that p^2, and therefore p, are even. But if p were even, it could be expressed as $2r$, in which case p^2 would be $4r^2$. But if $4r^2 = 2q^2$, $q^2 = 2r^2$, whence by the same argument as above, q^2, and therefore q, would have to be even. But then p and q would both have a common factor, contrary to our hypothesis. Therefore there are no two natural numbers, p and q, with no common factor, and such that $p^2 = 2q^2$; and therefore no rational number p/q equal to the square root of 2 [10].

It follows that the set of rational numbers, although dense, has a gap at $\sqrt{2}$, and therefore that if public time has only order-type η, there is no instant at which the freely falling body described at the beginning of this section passes the point 32·2 feet below where it began to fall. This would be counter-intuitive. Just as the Pythagoreans, starting with only commensurable magnitudes in two perpendicular directions, were compelled to admit incommensurable magnitudes in the hypoteneuse, so we, starting with only commensurable distances in one dimension of space and some constant accelerations, are compelled to admit incommensurable magnitudes in time. Quite apart from this metrical argument, it would be topologically unacceptable for there to be gaps in the succession (in the weak sense, based on the relation *after*, not *next after*) of temporal instants. For then the instantaneous present would not be

[10]   The discovery was made by one of the Pythagoreans, perhaps Hippasus of Metapontum. The Pythagoreans' interest in music and harmonies would have led them to consider whether any root that was not a whole number could be expressed as a rational number; e.g. whether by repeatedly taking the dominant we could ever get back to the original note, some octaves higher; that is, whether $(3/2)^n = 2^m$ exactly, for any n, m. In fact, we can never achieve this exactly; Bach's "Even-tempered Clavichord" was a final recognition of the fact. But K. von Fritz, "The Discovery of Incommensurability by Hippasus of Metapontum", *Annals of Mathematics*, XLVI (1945), pp. 242–64, suggests that it was a consideration of the geometrical properties of the regular pentagram that led to the discovery.

always the dividing line between past and future, but sometimes, while the terms past and future would be applicable, there would be no instant that was the present instant. This would be highly counter-intuitive. A "gappy" time would be even worse than discrete time: "*nec aliquo tempore*", said St Augustine, "*non erat tempus*" [11]. We may call this St Augustine's first doctrine of the ever-present present. It expresses our intuitive belief that there are no gaps in time, that there is no "crack between today and tomorrow" which something might disappear into, and in particular that there always is an instantaneous present dividing the future from the past. But such a division is a Dedekind "cut" (see § 7, p. 36), and if we accept it as a condition of temporal discourse that not only the terms 'past' and 'future' must be applicable, but also the instantaneous 'present', then it follows that temporal instants have the order-type of the real numbers, the order-type θ. St Augustine's first doctrine of the ever-present present enables us to build up a temporal continuum from temporal instants by means of Dedekind cuts, just as his doctrine of the ever-shrinking present enabled us to construct temporal instants by means of Cantorianly nested temporal intervals (see § 4).

[11] *Confessions*, bk XI, ch. XIII, xxvi.

§ 7
The topology of time

The conclusion that time is a continuum with the order-type θ, the order-type of the real numbers, is so important that we need to consider the argument further and make explicit the premisses on which it is based.

The real numbers satisfy the following conditions [1]:

(*i*) They are completely ordered; that is, given any two different real numbers, one of them is larger than the other. They can therefore all be ordered in size.

(*ii*) They are dense; that is, given any two different real numbers, there is a third one between them.

(*iii*) They are linear; that is, they form a *one*-dimensional continuum. This can be formally defined by saying that between any two real numbers there is a real number which is a member of a certain denumerable subclass. For the real numbers, the denumerable subclass is provided by the rational numbers, which turn out, surprisingly enough, to be denumerable; for ideal instants, the denumerable subclass is provided by a sufficiently fine mesh of markers marking off sufficiently small isochronous intervals.

[1]   E. V. Huntington, *The Continuum* (2nd ed. Cambridge, Mass., 1917), ch. V, esp. § 54.

(*iv*) They can be divided into Dedekind "cuts"; that is, if we divide all the real numbers in an interval into two (non-empty) classes so that every member of the one is smaller than every member of the other, then there is a real number "dividing" these classes, i.e. such that every real number smaller than it belongs to the one class, and every real number larger than it belongs to the other.

The same conditions hold for temporal instants:

(*i*) Temporal instants are completely ordered. That is, given any two different instants (at the same place), one of them comes after the other.

This is usually taken for granted, as by Hume (see § 6, p. 30), but needs the one vital proviso that the dates considered are all dates of events at the same place. In relativity theory, where we consider the dating of distant events, the ordering is no longer a complete one. The problem there involves the metric we need to ascribe to certain durations of time, and we must defer discussion until later (§§ 44, 46). Meanwhile, we should not be worried by the need for the proviso. If, as we maintain, public time is based on our private experience of time, in virtue of the intersubjective invariance of temporal order, there is an implicit restriction to events in the locality of observers all of whom can communicate with one another without delay. Although we may go on to ascribe dates in public time to distant events, we are *a*scribing them, not experiencing them directly. There is an element of artifice therefore in our dating of distant events. It is only the temporal order of·our own experience which is given us. And this order is complete.

Some philosophers have disputed it. Bradley suggested that time need not be "connected", as the topologists say, but there might be several different times, unrelated to one another [2] – much as if when we tell fairy stories and begin "Once upon a time", the story's fictional time, although unconnected with our time, was nevertheless a real time too, and the story true. But if there were really events in a time unconnected with ours, how could we know of them? It is all right in the case of fiction, because, being fiction, there is not supposed to be any criterion of truth independent of the storyteller's telling. We cannot say that Trollope has given us an unreliable chronicle of Barset, or that Conan Doyle is inaccurate about Sherlock Holmes, just because the author of a work of fiction is the only authority there is. We cannot convict a storyteller of factual error, for there are no facts for him to be in error about. And, *per contra*, if a man tells of events in a time inaccessible to us, we know that he is relating fables, not telling the truth. This for two reasons, one an obvious epistemological one, the other a logical and much more profound one. Against Bradley the verification principle is obviously effective. We can properly ask how he would know of events in another time; if he can know of them, he can know when they occurred in comparison to the other events of his life. If he cannot know of them, we need not take any cognizance of them either.

[2] F. H. Bradley, *Appearance and Reality* (Oxford, 1930), ch. 18, pp. 186–8, esp. 188.

The connectedness of time, however, goes deeper than this. For, as we have seen (§ 2, pp. 7–8), not only is time a necessary concomitant of my existing as a conscious being, but some relation to my existence as a conscious being is a necessary condition of time's being time. This is one of the most fundamental ways in which time differs from space. We could conceive of a space (indeed mathematicians often do) – although we could not know anything about it – that was totally unrelated to us, one in which we could not locate ourselves at all; for we can conceive of a mind that is not located in space, and can fairly easily take a God's eye view of space without locating ourselves within it. But we cannot similarly divorce ourselves from time, or abstract time from all connexion with ourselves [3]. It is a dim sense of this that underlies McTaggart's argument for the Unreality of Time [4], and a failure to appreciate it that vitiates the doctrine of eternity put forward by St Augustine and Boethius (see § 55, pp. 303–4). It is a point that will recur again and again, for it is one of the most fundamental ways in which time differs from space, and it gives time a unity deeper than anything we can establish for space (§ 9, pp. 58–9, § 44, pp. 216–17, § 52, pp. 279–81).

Although a totally disconnected time is conceptually impossible, it does not follow that time must be completely ordered. Prior talks of "branching time" [5] and Swinburne has suggested a temporary bifurcation of time followed by its ultimate reunification [6]. Prior's usage is idiosyncratic. He is talking about different courses of events rather than different courses of time as such, because he feels that time not embodied in events is too tenuous to talk about. But that is not to say that the tenuous time we do talk about is not completely ordered. Swinburne is denying that it necessarily is. His suggestion gains its plausibility from the contrast between public time and private time. Since public time is a construct from private time, it is *prima facie* conceivable that mankind should be for a period divided into two tribes each with its own, different, inter-subjective time, which later merge with just one unified public time. But the more fully we attempt to conceive it, the less conceivable it becomes. Essentially, in order to reach the conclusion that there were two times, the two tribes have to establish some identity of reference – they were talking about the same thing – and some discrepancy of description – they are saying different things about it. But the former is fragile. There is always the alternative, that time was not divided, and that the two groups were talking about things that were similar

[3]   Compare Kant, *Critique of Pure Reason* (London, 1929), A34/B50–1. Whereas space is only the form of our outer sense, time is the form of our inner sense, and "All appearances whatever ... necessarily stand in time-relations".
[4]   J. M. E. McTaggart, *The Nature of Existence* (Cambridge, 1927), ch. 32, §§ 305–6, vol. II, pp. 9–10. M. A. E. Dummett, "A Defence of McTaggart's Proof of the Unreality of Time", *Philosophical Review*, LXIX (1960), pp. 497–505, esp. pp. 502–3.
[5]   A. N. Prior, *Past, Present and Future* (Oxford, 1967), p. 50.
[6]   R. G. Swinburne, "Times", *Analysis*, XXV (1964–5), pp. 185–191; "Conditions of Bitemporality", *Analysis*, XXVI (1965–6), pp. 47–50; and *Space and Time* (London, 1968), ch. 10.

but not numerically identical. Swinburne protests against massive reduplication *praeter necessitatem*, but at the very least, it is open to us to claim that the undividedness of time constitutes such a necessity, and beyond this, Swinburne must provide some features in his account – some discrepancies of description – that will enable us to argue, as Hollis does [7], the non-identity of discernibles.

We can make another criticism of the Swinburne myth. It seeks to avoid inconsistency by keeping the two tribes separate, and to avoid vacuity by making them ultimately merge. But then the argument for there having been a temporal division depends on memory alone, and cannot be checked up on in any further way. If, however, the two tribes were to be able to communicate, then they would both have to have the same direction of time (see § 8, pp. 45–7), and hence the same time order, and we should be able to construct a unified public time, just as we do at present, with things as they actually are; and any feature that told against the unity of time – Hollis sketches some that might be envisaged [8] – would be construed as a simple inconsistency which told equally against the plausibility of the whole story [9]. Swinburne therefore must keep his two tribes separate during his bitemporal period, and only allow them to interact after the two times have been reunified. But then his whole argument becomes thin. Memory is fallible. Only coherent memories are accounted reliable. And the memories of Swinburne's tribes necessarily do not cohere.

(*ii*) Temporal instants are dense. Given any two different instants, we can always conceive of an event occurring at an instant after the earlier one and before the later one.

We have argued this point enough in the previous section.

(*iii*) Temporal instants are linear: that is, there is only one dimension of time, and its ordering is Archimedean.

We are seldom tempted to think of time as two- or more-dimensional in the ordinary sense, although Eddington has produced an argument for the three-dimensionality of space which, if valid, would prove also the two-dimensionality of time [10], and Dunne has argued for an infinite number of temporal dimensions [11]. The main argument for there being only one dimension of time must be deferred until we have elucidated what we mean by dimension (see § 36, pp. 178–9); but in the ordinary way of reckoning, it would be incompatible with the complete ordering of time, condition (*i*), to have more

[7]   Martin Hollis, "Box and Cox", *Philosophy*, XLII (1967), pp. 76–8; "Time and Spaces", *Mind*, LXXVI (1967), pp. 534–5.
[8]   Martin Hollis, "Time and Spaces", pp. 528 ff.
[9]   A. M. Quinton, "Spaces and Times", *Philosophy*, XXXVII (1962), pp. 144–6.
[10]   A. S. Eddington, *New Pathways in Science* (Cambridge, 1935), ch. XII, §§ iv, v, pp. 267–77; reprinted in James R. Newman, *The World of Mathematics* (New York, 1956), vol. III, pp. 1567–73. Eddington himself does not draw this conclusion.
[11]   J. W. Dunne, *An Experiment with Time* (3rd ed. London, 1934), ch. XX; and *The Serial Universe* (London, 1934).

than one dimension. For to have two or more dimensions is to have two or more metrical relations between points, and *a fortiori* two or more orderings. However, we can collapse a two- or more-dimensional manifold onto a one-dimensional ordering by an artifice: for example, we could say that the points on a plane (x_1, y_1), (x_2, y_2), ..., etc., were to be ordered by the following rule: if $x_1 < x_2$, then (x_1, y_1) comes before (x_2, y_2) and vice versa; if $x_1 = x_2$, then according to whether $y_1 < y_2$ or $y_2 < y_1$, (x_1, y_1) comes before (x_2, y_2) or vice versa. Such an ordering is called non-Archimedean, and sometimes we are tempted to ascribe a non-Archimedean order to our temporal experience to take account of our dreams or our moments of exaltation or depression. "Time has no meaning while the gypsies sing", I live a whole other life of strange experience during a single cadence of their song. In the moment of truth, when I see myself as I really am, I die a thousand deaths. One American story of the Civil War obtains its whole effect by narrating *in extenso* how a condemned man thought he had escaped and all that he thought he was doing in the interval between his being dropped from the gallows and the noose's actually breaking his neck [12].

Condition (*iii*) requires us to reject such accounts as, at best, metaphorical or fictional, at worst false. We are to reject such events or experiences as being "outside time" (outside public time, that is), thereby securing to public time a linear ordering. There is some difficulty, however, in applying the stipulation: whereas with the real numbers, the rational numbers constitute a suitably denumerable subclass, we cannot work a similar subclass of all temporal instants. Instead we make an approach in two stages. We first allow it as a positive indication that two *soi-disant* events in a putative temporal order are ordered in public time if some publicly observable event – e.g. the chiming of a clock – occurs after the one and before the other. If in my brown study, I can say which of my thoughts occurred before the clock struck three and which after, then we allow that my thoughts were in time, even if far away from the topic under discussion. But this test is incomplete. The college clock strikes only every quarter, and therefore many events in my consciousness are not separated by its chimes. But if it chimed more frequently, they might be. And if, no matter how frequent was the succession of public events, there could never be one of them that came after one and before the other *soi-disant* events, then we would say that the putative time order of the *soi-disant* events was not that of real time at all. Thus the second stage is to consider a denumerable set of denumerable sets of public events closer and closer together. We imagine an infinite set of clocks, each ticking away *ad infinitum*, each ticking faster than the previous one. It is then a sufficient and necessary condition of

[12]   Ambrose Bierce, "An Occurrence at Owl Creek Bridge", in *Tales of Soldiers and Civilians* (New York, 1891), pp. 21–39; and in *In the Midst of Life* (London, 1919), pp. 31–41. Compare the non-Archimedean time adopted by C. S. Lewis in his Narnia stories, e.g. *The Lion, the Witch and the Wardrobe* (London, 1959, paperback ed.), pp. 27, 48–9; but see note [15] below.

two *soi-disant* events being in a real temporal order that they should be separated by some tick of some such clock. And the denumerable set of denumerable sets of ticks will comprise itself a denumerable number of ticks; for we can number them diagonally, just as we do the rational numbers.

Condition (*iii*) thus is to be construed not as a condition on any conceivable concept of time, but only on our concept of real, public time. We concede to the speculative metaphysicians that we might have a non-Archimedean ordering of time, in which we "inject" dreams [13] or, *à la* Bradley [14], novels into our basic time, and keep order according to the following rule: if we are comparing dates of events in different books, we go by date of publication; if we are comparing dates of events in the same book, we go by the internal dating system of that book. Such a rule would give us a complete ordering, but a non-Archimedean one. *But* to preserve its non-Archimedean status, it is essential that the dreams and books should be internally isolated from current affairs. Apart from the date on the title page, or the instant of falling asleep, there must be no other connexion between events inside the book or the dream and ordinary events. And this in itself is a mark of unreality. If a story were real, between any two chapters the hero could have written a letter to me or someone, asking advice on what his next step should be. If, in principle, between any two instants of dream time, I could date an intersubjectively datable event of public experience, then it would be conceded that dream time could be fitted in with public time, and was, therefore, real [15].

Condition (*iii*) is in practice guaranteed by our use of clocks to date events in public time. Deferring until § 12 the metric question of whether clocks mark out isochronous intervals, we can describe the successive ticks of a clock topologically as being of order-type $\omega^* + \omega$, and in principle we set no upper limit to the fineness of discrimination of clocks. Given any two instants, there could always be some clock that would tick between them. Therefore, if we order our public time by clocks, we have conceded condition (*iii*).

The fourth condition is the one we have just argued for in the previous section. It requires that:

(*iv*) Temporal instants can always be divided into Dedekind cuts, by the application of the terms 'past', 'present' (in its instantaneous sense) and 'future'; that is, besides a past and a future, there must always be a present.

[13]   See, for example, A. M. Quinton, "Spaces and Times", *Philosophy*, XXXVII (1962), pp. 130–47, esp. § 6, pp. 144–7.
[14]   *Appearance and Reality* (Oxford, 1930), ch. 18, pp. 186–8, esp. 188.
[15]   In C. S. Lewis's *The Lion, the Witch and the Wardrobe* (London, 1959) events occur in Narnia even when no children are there. We may ask when, in human time, the events discovered on p. 56 had actually occurred – at what human time the children would have had to have ventured through the wardrobe in order to have been able to intervene, and perhaps rescue the Faun. Once this question is answerable, the time ordering of the children's experience ceases to be non-Archimedean.

There are some purely mathematical difficulties. They are not difficulties about the concept of time, but about any continuum. Some, like Zeno's, have been decisively answered, and are only raised by those who do not understand the mathematicians' concept of the continuum. Others are less easily brushed aside. They are based on the fact that in order to complete our account of real numbers, and to open the way for classical analysis, we need to use "impredicative definitions". For to establish the fundamental "least upper bound" theorem, we need to take Dedekind cuts not only of the rational numbers but of the real numbers as well, and require that a real number may be defined by a Dedekind cut of all the real numbers, itself presumably included. Some philosophers find this objectionable. I do not. Nor do most mathematicians. Impredicative definitions, although difficult to swallow, are not circular, certainly not viciously so. In any case, the concept of the continuum stands or falls along with the main edifice of classical mathematics [16], and if we allow any ordinary mathematics, we must allow the continuum as a viable concept.

Mathematical objections apart, the argument, as stated, is suspect. The use of the word 'always' seems to beg the question. An objector could argue that "There always is a present" is a tautology; and that to argue that therefore time is continuous depends on assuming a continuity in the 'always' which one is trying to establish for all time generally. The argument, however, is not circular. The word 'always' is being used as a modal operator, not a temporal one. What we want to say could be rephrased, aseptically, by saying "In no case could we conceive of time past and time future without there being an instant – the present instant – dividing them." The three terms, past, present and future, are either all applicable (though to different intervals and instants) or all inapplicable: but where they are all inapplicable, we are not speaking of time at all. Therefore, it is a necessary condition of our speaking in a temporal context, that the term 'present instant' is applicable; that is, that there is some instant that can be correctly described as the present instant. But if this is granted, then it follows that in all cases where we talk of time at all, we ascribe to it the order-type θ of the real numbers.

It is therefore legitimate to use the formulae of Galilean kinematics (as at the beginning of the previous section, p. 29) and of all mechanics, which implicitly assume the continuity of time. So too the more fundamental assumption of § 2 is justified. Public time is like private time in being of order-type θ, although what is interpersonally invariant is only of order-type η. We are right therefore to ask the question 'How much?' of public time, just as we do each of our own subjective, private time; but before we can go on to answer this question by setting up a metric for public time, there are two further properties we need to establish. Time is not cyclic. And this topological property follows from an even more fundamental one – that time has a direction. Some of the

[16]   But see Hermann Weyl, *Das Kontinuum* (Leipzig, 1918), for an (unsuccessful, as I think) attempt to reconstruct classical mathematics without impredicative definitions.

arguments used will depend on the fact that we are rational agents, and not merely sentient beings, and thus anticipate the considerations of §§ 51–54. Even without them we should still be able to argue that time, merely as the concomitant of consciousness, was directed and non-cyclic: passengers have some idea of the order of the places passed through – but much less sharp or reliable than that of the driver.

§ 8
The direction of time

Time always has a direction. If we can talk of time at all, we can talk of 'before' and 'after', and these are essentially asymmetrical relations. If we could not order events according as they occurred before or after one another, we could not regard them as really temporal. Directionless time is not time at all.

Everybody believes that time has a direction, but many allow themselves to be confused about it. Some reduce it to the trivial contention that any ordering relation is necessarily asymmetrical (or antisymmetrical), and make out that the difference between before and after is no more than that between north and south. Others confuse time with change, and since changes often can be reversed, argue that time must be reversible too. Many scientists get into great difficulties because, partly under the influence of Plato and the geometers, partly in view of the so-called time-reversibility of classical physics (see further § 49) or the conservation of TCP [1], they form a static picture of the world in

[1] For a readable account, see Martin Gardner, "Can Time go Backward?", *Scientific American* (January, 1967), pp. 98–108. Many scientists have suggested that it might be, or is, possible for some processes or particles to run backwards in time; for example, G. N. Lewis, in *Nature*, CXVII (1926), p. 235, and in *Science*, LXXI (1930), pp. 569–77; E. C. G. Stückelberg, "Remarque à propos de la création de paires de particules en théorie de relativité", *Helvetica Physica Acta*, XIV (1941), pp. 588–94, and "La méchanique du point matériel en

which there is no given direction of time, and then try to put one in. But that too is a mistake. The direction of time is given us from the start. We may, for certain scientific purposes, leave it out. But we could not, and fortunately need not, ever have to look for it to put it in again: and while the Second Law of Thermodynamics is of great importance, if it should fail, or be interpreted some other way, it would not mean that time might then run backwards.

When philosophers raise the possibility of time going backwards, they may mean several things. They may envisage one man having a private time going in the opposite direction to that of everybody else; or they may be wondering whether Time Machine stories are conceptually possible; or they may be concerned to interpret some anomalous entities in particle physics, in view of the principle of TCP conservation; or they may be unravelling the ties between the direction of time and other concepts of common sense or natural science.

We shall argue first that everybody's private time must have the same direction – that it is impossible for one man to have a private time going in the opposite direction to that of everybody else. For this we shall use a communication argument. Communication arguments are of great power in accounting for our conceptual structure of time and space. We shall use them not only here to account for the uniform direction of time, but to provide a transcendental deduction of the Lorentz transformation (see § 44), to show why space has to have three dimensions (see § 48, pp. 244–6), and why we prefer to impose a Euclidean rather than any other geometry on our spatial experience (see § 38, p. 183 and n. 1).

Considering himself in isolation, a philosopher may allow that time is involved in change, and that it is a necessary condition of experience. But it might be a purely subjective matter. Each agent and each sentient being might have his own time. After all, our sense of the passage of time, our experience of duration, is private, and can vary very much from person to person: and we might wonder whether our temporal experience might not differ in other, more radical, ways from person to person, and in particular, whether its direction might not be different for different people. Could we not, in a looking-glass land, meet people whose temporal experience ran in the opposite direction to ours? We know that in practice we do not, but it is not immediately

---

théorie de relativité et en théorie des quanta", *Helvetica Physica Acta*, XV (1942), pp. 23–37; R. P. Feynman, "The Theory of Positrons", *Physical Review*, LXXVI (1949), pp. 749–59. I do not discuss the scientific merits of their proposals – in particular I do not discuss the fascinating suggestion of J. A. Wheeler that all electrons are qualitatively identical because they are all the same electron threading forwards and backwards an enormous number of times – because the general arguments I shall adduce in this section would always suffice to tilt the balance of plausibility against such proposals. In particular, we should always impute discrepancies to non-conservation of charge or parity, rather than to a reversal of time; and indeed, I would far sooner abandon Time-Charge-Parity conservation altogether than envisage the possibility of time going backwards.

obvious whether this is just a brute empirical fact, or a conceptual necessity of some sort. But if we consider more carefully what is involved in two beings communicating with each other, we can see that it is a necessary condition of their being able to communicate that they should experience time going in the same direction. A uniform direction of time is therefore an essential condition of intersubjective experience.

There are three arguments. The first concerns the causal means of communication. We communicate by interacting on one another. I speak, and produce vibrations in the air, which spread out and impinge on your ear drums, and make you have some sort of auditory experience. Or I write, and make visible marks on paper, which later you see, and construe your visual experience as a message from me. Or I might tap out or flash messages in Morse Code, or by semaphore or by teleprinter or by flying certain flags. In every case the sender *causes* the recipient to have certain experiences which he takes as signs of the sender's intentions. But if two people had time orders that ran in opposite directions, they could never send and receive messages. If either thought that he was sending a message to the other and that the other was receiving it, then the other would not construe it as being received, but as, again, being sent, by him. For, according to the one in his time order, as time progressed, the causal disturbance that constituted the message would approach the other: but according to the other, since his time order was reversed, the causal disturbance would appear not to be approaching him but to be going away from him; so he would construe it, if he construed it as a message at all, as a message which he himself had sent out. So, too, for any message that the one thought he was receiving from the other, the other would also think that he was receiving from the one. Any message that was in the out-tray of the one would be in the out-tray of the other, and any message that was in the in-tray of the one would be in the in-tray of the other. Messages would never get from the out-tray of the one to the in-tray of the other, or vice versa, and so on this score alone communication would break down.

Communication would break down, secondly, because of the interchange of efficient – or, better, "causal" – causes and final causes. Messages are composed of conventional signs, effected, or caused, by the sender, with the intention of their being construed in a certain way, and taken by the recipient as being intended to be construed in that way. Intentions and signs depend on the direction of time, as well as on causes and effects. Each of the two beings, therefore, would observe the actions of the other in reversed time-order but would construe it in his own. If I felt a fly settle on my nose and scratched my nose in consequence, you would see me put my finger to my nose, and the fly thereupon alight, and you would construe my action not as caused by the fly's alighting but as causing it and being intended so to do. With flies on noses, the resulting account may be bizarre, but need not be unintelligible. With communication, however, the difference between causal antecedents and intended consequences is crucial. I regard some experience or some thing as a sign if I think it is caused by some rational agent who intended it to be taken as a sign. So far as communication is concerned, we are indeterminists [2]. Messages are taken to be messages only if they are taken to be caused not by natural phenomena but by beings who can initiate causal disturbances, and alter the course of events from what it would otherwise have been, and have done so in order to give us to understand what was in their minds. If we reversed the order of time we should have to construe the signs composing the message as the causal antecedents of the agent's behaviour, not the intended consequence. Speech would appear as a chance build-up of sound waves, suddenly phasing one another out and collapsing into silence, leaving the "speaker" looking attentive, no doubt startled by the sudden noise. Writing would appear to be erasing. If we regarded the "writer" as an agent at all, he would appear to be a waste-paper merchant, making soiled paper as good as new. More likely we should regard his erasure as some natural phenomenon involving loss of structure, like burning, or overdeveloping a photographic plate [3].

The third argument is based not on the means of communication or their interpretation but on the logic of what is communicated. Even if we could communicate telepathically, without any causal interaction, we should still be unable to communicate successfully unless we had the same order of time in common. For although, conceivably, we might be able to understand a single message backwards, we could not engage in any two-way conversation. We could not ask questions and get answers, and we could not argue with each other.

We might be able to unravel a single message backwards. One can, after all, read somebody else's newspaper backwards, in a train: in one history book some chapters have been written starting at the end and working backwards to the beginning [4]. If a man put forward a concerted argument, we might be able to understand it, provided we translated his 'therefores' into 'becauses', his 'sinces' into 'sos', and his 'ifs' into 'only ifs.' We could understand logic backwards. But we could not have any sort of dialogue. If I were to ask a question, and he to answer it, either he would answer it before he heard it asked, or I would have asked it already having heard what the answer was. The latter supposition is absurd. The whole point of asking questions is that one does not know what the answer is, and hopes that one's respondent does. If one hears the answer before asking the question, the question is otiose. "What is the date of the battle of Hastings?" I ask. My respondent hears the question and answers it, a few seconds later by his time, "1066". Therefore,

[2] See further F. J. Crosson and K. M. Sayre (eds.), *Philosophy and Cybernetics* (Notre Dame, 1967), pp. 26–9; and J. R. Lucas, *The Freedom of the Will* (Oxford, 1970), § 11, pp. 63–4, § 16, pp. 88–9.
[3] See further Norbert Wiener, *Cybernetics*, 1st ed. (New York, 1949), pp. 44–5, 2nd ed. (1961), pp. 34–5.
[4] B. H. Sumner, *Survey of Russian History* (2nd ed. London, 1947), chs. II and III.

by my time, I had heard "1066" before asking my question. Therefore my question could not be a question but a comment "1066" – "That is the date of the battle of Hastings". I have to translate "What?" by "That", "Who?" by "He", "Where?" by "There". But whereas it is just possible to translate my respondent's terms into my own, it is incoherent to translate my own into other, different, ones of my own. If I can only comment, and not ask questions, it is no longer a conversation I am conducting, but only an interjection of scholiast's notes.

The former supposition, that he would have answered my question before he had heard it asked, is equally absurd, although less obviously so. Questions can sometimes be anticipated, but not always so. It would destroy epistemological parity between speakers if one of them knew in advance all the questions that were going to occur to the other's mind. Once again, the dialogue would have collapsed into a monologue. He would be telling him things, but not conversing, not talking with him. To talk *with* demands parity of epistemological status. Even if one knows all the answers, one cannot know always which answer is wanted in advance of the questioner's indication of his ignorance. It would also destroy parity of esteem. Not only would you have to see through me, in order always to answer my questions before you heard them asked, but you would soon come to despise me, if I appeared to keep on asking questions just after having been told the answers. Moreover, even if single questions and answers could be transposed, often we have to ask a whole series of supplementary questions to elucidate the first answer we are given. But to answer the supplementaries before giving the answer they were designed to elucidate would be a confession of incompetence: for if one knew they were going to be asked, one should phrase one's main answer better, so as to avoid ambiguity or difficulty.

Not only question and answer but also argument would be impossible between beings whose time orders went in opposite directions. I cannot argue against your contention until you have stated it, and you cannot counter my objections until you have heard them. Although occasionally we may, with Aristotle and Newman, anticipate objections, to do so generally is to advance a monologue rather than to engage in a genuine dialectical argument. Real argument is a meeting of minds. But minds cannot meet if from every encounter they are necessarily going in different directions. They must be able to interact as a matter of logic, as well as a causal condition of the possibility of communication. And therefore if two beings are to regard each other as communicators, they must both have the same direction of time. It is a logical as well as a causal prerequisite.

The direction of time ties in with other parts of our conceptual structure of common sense and natural science in five, interrelated ways:

(i) We can alter the future, but not the past.

(ii) We can remember the past, but not the future.

(iii) Efficient (or better, causal) causes are antecedent in time; final causes are subsequent in time.

(iv) Entropy increases.

(v) Biological organization increases.

The first of these concerns our status as agents, and involves our notion of cause; the second our status as observers, and involves our notion of knowledge; the third reveals the connexion between time and explanation; the fourth indicates the chief part played by the direction of time in physical phenomena; and the fifth indicates the contrary part it plays in biology. They are interconnected.

We can alter the future. If we could not, we should not be agents. To be an agent is to be able to form intentions and make plans and to carry at least some of them into effect. What the future will be is at least in part up to us. We do not merely predict what it will be, but decide what it shall be. Although determinists, fatalists, and other faulty philosophers make out that our decisions are just as much predictable as anything else, and that the future is entirely determined by antecedent conditions and not at all open to our intervention or influence, we do not really believe them, and, if we do, do not believe ourselves to be the beings we know ourselves to be.

We cannot alter the past. As Agathon said

μόνου γὰρ αὐτοῦ καὶ θεὸς στερίσχεται,
ἀγένητα ποιεῖν ἅσσ᾽ ἂν ᾖ πεπραγμένα.

Of this alone even God is deprived – the power of making not to have happened those things that have been done [5].

This is not to say that we cannot ever undo what we have done. We often can. Change, unlike time, can be reversible, and having made some change, we may be able to change things back again to how they were at the outset. We can, in that sense, sometimes undo the past. But we cannot cancel it. If we change and then change back, although the last state will be the same as the first, it will not be the case that there never had been a change. There was a change. Only, there was another change too, which undid the effect of the first change. Although the final state of affairs is as if there had never been any change, it is not true that there never was any change or counterchange.

Not every change can be reversed. If we cut down a tree we cannot set it up again, nor bring to life organisms we have killed. Particularly in human

[5]   Aristotle, *Nicomachean Ethics*, bk VI, 2, 6, 1139b 10, 11. Cf. Aquinas, *Summa contra Gentiles*, II, 25, 1023: "Deus non potest facere quod praeteritum non fuerit"; and *Protagoras*, 324b: "οὐ γὰρ ἂν τό γε πραχθὲν ἀγένητον θείη". For a contrary view see M. A. E. Dummett, "Can an Effect Precede its Cause?", *Proceedings of the Aristotelian Society, Supplementary Volume*, XXVIII (1954), pp. 27–44, and "Bringing About the Past", *Philosophical Review*, LXXIII (1964), pp. 338–59.

affairs, where memories matter, we cannot ever completely undo the past. For deeds are known, usually by others, necessarily by ourselves; and since knowledge is always important, the logical unalterability of the past becomes a contingent irreversiblity of human affairs. We can never put the clock back, because, however faithfully we recreate the situation before the deluge, people will remember the rain, and therefore will look on everything with water-mindful eyes. We may forgive, and in minor matters forget; but so long as we have memories, the past not only cannot be altered, but can never be altogether expunged.

Memory is indispensable to knowledge. If we did not remember anything, we could not know anything. There are many different sorts of memory, just as there are many different sorts of knowledge. But the different sorts of memory, like the different sorts of knowledge, are connected; and although partial amnesia in respect of only one facet of memory is a possible – and common – affliction, total amnesia in any one respect would destroy memory in other respects too. Language would be impossible without some recognition of faces or places or things, and, equally, without some recollection of one's own actions, and some power of recalling what one had been told. If I am to know anything, I must know who I am, and therefore who I have been and what I have done. Therefore I must remember if I am to be a responsible agent at all, and if I am to be a conscious being at all. What I know cannot be other than what it is. It must be unalterable. If, therefore, there are agents who exist in a world in which not everything is unalterable (else they could not be agents), who can know things, and know who they are, then there must be some things that are no longer alterable which go to constitute each agent as the particular agent he is – as the doer of the deeds that he has done. If knowledge was only of Platonic timeless truths, then there would be no ground for distinguishing one knower from another. Everything the one knew, the other could know too. If they are to be different knowers, they must know different things. So far as impersonal – third-personal – knowledge is concerned, it could never be necessarily the case that their knowledge was different, and so we could never be sure that different persons would continue to be different. Only personal – first-personal – knowledge will provide a satisfactory ground of differentiation between knowers. And what is primarily first-personal is agency – what I must know, and am the prime authority for others to know of it, and need the first person singular to tell them, is what I am going to do and what I have done. On the former I can change my mind. My intentions are peculiarly my own, but they do not constitute me the person that I am – rather it is for me to make them come into effect or not. My actions are, more than my intentions, public property; they are, once done, out of my control and identify me, and go far to constitute me, as the person who did them. And it is my having done them, at least some of them, that I must know if I am to know anything at all.

We cannot remember (or "pre-member") the future. If we could, it would not be alterable, and we could not decide what we were going to make of it. We can, within limits, decide the future – but we can always revoke our decision until it is put into effect. We can also, in part, predict the future, by the same methods as we can also retrodict the past, but more essentially. It is not essential to be able to retrodict, because I can remember; but it is essential to be able, at least to a limited extent, to predict, in order that I may control. Where predictions are absolutely certain, however, they are felt to limit the range of choice open to us; and if every event in the future were, as the determinists assert, predictable, we should feel both that our status as agents had been subverted, and that the genuineness of time had been destroyed. We should picture a block universe, in which the future was already fixed, already there, and in principle no different from the past. It is deeply counter-intuitive. Our intuitive sense of the passage of time is a passage from aspiration to achievement, from potentiality to actuality, from uncertainty to knowledge.

Time Machine stories differ in how much of our normal complement of knowledge and abilities is taken back into the past or forward into the future. Often we are invited to imagine ourselves witnessing past events. Provided we are inactive and invisible, no paradox need ensue. If we are merely passive spectators, we are not altering the unalterable past; and provided we cannot be seen, heard or felt by any of the actors of the events, our presence makes no difference, and the purity of the past is preserved untouched. If we voyage forward into the future there are greater difficulties. What shall we see? If it is "already there" for us to see, then the future is fixed, and we are wrong in our claim that it is alterable. Our whole conceptual structure is in the same jeopardy if precognition is possible as if determinism were true. It is very important, not only for reasons of modesty, that I should not be able to use a Time Machine to go into a public library and read my own biography. It would be all right for me in the future to be visible and active, and go round giving people the benefit of my greater wisdom, in order that they should avoid the disasters that I foresaw: but only on condition that it is merely a hypothetical future, a might-be, not a will-be. For only then could I come back to the present and continue to be an agent at all. I can voyage through many potential futures, and come back a wiser man; but let the future once be fixed, and it forecloses my liberty of action altogether, and I cease to be a man, and am only a passive spectator, fatalistically observing the story of my life [6].

Causal and final causes are both time-directed. We explain a phenomenon causally by citing some earlier condition from which, in conjunction with some law of nature and other standing conditions, the phenomenon could have been predicted. We give a final, or teleological, explanation of an action by citing some later condition towards which the action is, in virtue of various

[6]   Compare passage from H. Weyl, *Philosophy of Mathematics and Natural Science* (2nd ed. New York, 1963), p. 116, quoted above § 2, p. 13, n. 13.

laws of nature and other standing conditions, conducive. In causal explanations we refer back in time to obtain our explanation, in final ones forward. Both causal and final explanations are aspects of our being able to alter the future. Because we can alter the future, and want some future conditions to obtain rather than other ones, we implement those means that will bring about the ends we desire. We seek to discover the relation of cause and effect in order to do this, and learn it by remembering our various failures and successes in achieving what we want. The modern scientific ideal of explanation is made out to be an impersonal one, but it stems from an essentially personal activity [7]. I discover causes, not, as Hume thought, by passively observing invariable concomitances but by intervening and altering conditions myself, and trying and seeing what actually will happen. Causality is an empirical concept – that is, one depending on trial and experiment.

We can explain the end result by reference to the antecedent conditions that brought it about; and equally, we can give the end result as our reason for carrying out those means that would be effective in achieving it. It is only because the future is alterable that we can alter conditions and isolate causes, and, by trying and failing to obtain some combinations of conditions, acquire the sense of natural impossibility and natural necessity; and it is only because the future is alterable that we can explain our actions as being intended to alter it to some desired effect. The alterability of the future is thus a necessary, as well as a sufficient, condition of our being able to have the two, linked but contrasted, schemes of explanation.

Leibniz, who did not believe that time was real, minimized the distinction. He rejected the programme of banishing final causes from natural science, and claimed that a complete explanation in terms of efficient (i.e. causal) causes was compatible with one in terms of final causes, and, anticipating the principle of least action, offers a genuine scientific explanation in apparently final forms [8]. But maximum and minimum principles are not really teleological, and it is difficult to believe that a complete causal explanation could be compatible with a final, or any other sort of, explanation [9]. If we can distinguish between causal and final causes at all, we already have a criterion for distinguishing antecedent from subsequent states of affairs; and the fact that we have either stems from our ability to alter the future and remember the past.

The fourth and fifth time-directed principles, that entropy increases with time, and that biological organization increases with time, seem more peripheral to our conceptual structure, and also mutually opposed and to some extent

[7]   See R. G. Collingwood, *Metaphysics* (Oxford, 1940), ch. XXIX, XXXI, XXXII.
[8]   G. W. Leibniz, *Discourse of Metaphysics*, tr. P. G. Lucas and Leslie Grint (Manchester, 1953), §§ XVII–XXII, esp. XIX, XXII. See further H. Margenau, *The Nature of Physical Reality* (New York, 1950), § 19.11, pp. 422–5.
[9]   See J. R. Lucas, "Freedom and Prediction", *Proceedings of the Aristotelian Society, Supplementary Volume*, XLI (1967), pp. 170–1; and *The Freedom of the Will* (Oxford, 1970), § 10, pp. 47–50.

cancelling each other out. Whereas the choice of our future actions and the memory of our past ones are considerations immediately present to our consciousness which we can barely think away, and our schemata of explanation must play a large part in any rational account we give of our conceptual structure, the laws of physical and biological development are matters more of observation and discovery, which we might conceive, at least on a first consideration, to have been otherwise. A film run backwards portrays the sequence of visual sense experiences of an observer (not an agent) in a time-reversed world. In such a film show, apart from the evident incongruities of human behaviour, we should notice some other features that would suggest what had happened. Water would flow uphill, mixtures would separate, temperatures would diverge; plants would lift up their seeds from the ground, their fruits would shrink, petals would float in from the wind and reflower their blossoms, which in turn would shrink into buds, and the whole plant would gradually ungrow into its seed.

Such phenomena are evidently different from those we are familiar with, but it is not obvious that they are necessarily so. The very fact that we can run a film backwards and give an intelligible description of what we should see shows that it is not logically impossible for entropy or for biological organization to decrease instead of increase. Nevertheless, the actual directions of physical and biological development are not merely contingently concomitant with the direction of our intending and our remembering. The Second Law of Thermodynamics, that the entropy of an isolated system increases with time, means that each system tends to move towards a less differentiated state. A cup of tea cools down to room temperature (and the room warms up, very slightly); an ice cream melts (and the room cools down, again only very slightly). Whatever the initial temperature, it makes no difference in the end – it will be at room temperature. Even if we consider the slight change of room temperature, it will be all one whether we started with a small cup of boiling tea, a large bowl of warm water, or a large bowl of hot water together with some ice cream. Worse, it would make no difference if the ice cream had been boiling, the tea iced tea. The end result of all these different initial conditions would be the same. Different causes (under a suitable description of what a causal situation is) have the same effect. And this is, crudely, what it is for entropy to increase.

Let us now tabulate the four possible causal principles:

The same cause may have the same effect.
Different causes may have the same effect.
The same cause may have different effects.
Different causes may have different effects.

These principles are not fully determinate until we have specified criteria for sameness and difference: hence the 'may'. Leaving this important point aside, we note also that the propositions are not all mutually exclusive; in particular, the first and the last are normally complementary. But there is some incompatibility between the first and the third, and between the second and the fourth, in so far as the principles are being applied to the same range of situations in virtue of the same criteria. We therefore have three possibilities:

Same cause, same effect: different cause, different effect.
Different cause, same effect.
Same cause, different effect.

The first of these represents a completely determinist system, such as Newtonian mechanics, where, granted a complete specification of initial and final conditions, there is a one–one correspondence between them, and to each initial condition – cause – there is one and only one final condition – effect. The second represents the case described by the Second Law of Thermodynamics, where there is a many–one correspondence between causes and effects. The third represents the contrary case, in which there is a one–many correspondence between causes and effects. But whereas either the first or the second principle is compatible with our activity as agents, the third is not. For if we are to be able to alter the future, we have to be able to manipulate things and circumstances in order to bring about the desired state of affairs. And we can know what things to do only if we know some causal law, some generalization that if something of this sort happens, then something of the desired sort will happen. But such knowledge is only possible on condition that the same cause (something of this sort) generally produces the same effect (something of the desired sort). If the same cause was liable to produce different effects, without there being any further factor to determine which effect would in fact be produced; if sometimes the ice cream would get cold the tea get hot, and then, separate itself into tea and milk; and if at other times the ice cream would heat up and the tea cool down; then we could not control the course of events at all. We could not say that we could alter the future, because, whatever our actions, their consequences – if consequences they could be called – would be quite unpredictable. The future would diverge from what it was at present and had been in the past, but in totally unpredictable and improbable ways. The future would be different: that is all we should know about it. But that is too indefinite to be a guide for action. Therefore if the contrary of the Second Law of Thermodynamics were true, and entropy decreased with the passage of time, we could not as agents alter the future.

It remains formally possible that entropy neither increases nor decreases, but remains always constant: that is, that the same causes have the same effects, different causes different effects. It would be possible to operate in such a world, although for our, sub-Laplacian, intelligences it might be somewhat difficult. Our actual world, with its limited available information, makes fewer demands on our limited intellectual powers. Nor is it only a practical matter.

Explanations count as being genuinely explanatory only if what is tendered as an explanation is more intelligible, or intellectually acceptable, than that which it is intended to explain. It is a criticism of the completely determinist view of the universe that in its efforts to explain everything totally, it has succeeded in giving no real explanation at all for anything; for the explanations it offers are all causal explanations in terms of initial conditions, which are themselves liable to be as opaque as the states of affairs they were supposed to explain. The only case where an initial condition by itself is felt to explain a subsequent one is where it can be described more simply. The emission of light from a point source is intelligible in a way in which the focusing of light on, and its absorption at, a point source is not intelligible, just because in the former case the initial condition admits of a very simple description – one position occupied by a point source, everywhere else empty – whereas in the latter the description of any possible initial condition would be fiendishly complicated [10]. To count as an adequate explanation, the description of the initial condition must be more informative – contain more information according to some criterion of relevant information – than the description of that which is being explained. Therefore, if the explanation is to work, there must be a loss of information – that is, a gain of entropy – between the initial and subsequent conditions. If we are committed, as most of us are, to the view that causal explanations can be explanatory, we are committed, *pro tanto*, to some doctrine of the increase of entropy with time. It is not necessarily the precise doctrine of modern physics: for we have not formulated our criterion of relevant information; and it was apprehended by Aristotle, who first articulated the schema of explanation in terms of efficient causes, long before the rise of thermodynamics:

καὶ πάσχει δή τι ὑπὸ τοῦ χρόνου, καθάπερ καὶ λέγειν εἰώθαμεν ὅτι κατατήκει ὁ χρόνος, καὶ γηράσκει πάνθ᾽ ὑπὸ τοῦ χρόνου, καὶ ἐπιλανθάνεται διὰ τὸν χρόνον, ἀλλ᾽ οὐ μεμάθηκεν, οὐδὲ νέον γέγονεν οὐδὲ καλόν· φθορᾶς γὰρ αἴτιος καθ᾽ ἑαυτὸν μᾶλλον ὁ χρόνος. [11]

[10] K. R. Popper, *Nature*, CLXXVII (1956), p. 538, CLXXVIII (1956), p. 382, CLXXIX (1957), p. 1297, CLXXXI (1958), p. 402, argues from the phenomenon of radiation to the directedness of time, not as between one observer and another, but in physics itself. G. N. Lewis, *Science*, LXXI (1930), pp. 569–77, had argued that apparent directedness was due merely to the physicists' unjustified habit of discarding "advanced potentials" as physically feasible solutions of Maxwell's equations, and admitting only the "retarded potentials"; but Popper can cite non-electromagnetic phenomena where no question of "advanced potentials" can arise. See further discussion, Richard Schegel, *Nature*, CLXXVIII (1956), p. 381; E. L. Hill and A. Grünbaum, *Nature*, CLXXIX (1957), p. 1296; and R. C. Bosworth, *Nature*, CLXXXI (1958), p. 402; and G. J. Whitrow, *The Natural Philosophy of Time* (Edinburgh, 1961), ch. I, pp. 7–10.
[11] *Physics*, IV, 12, 221a 30: "A thing then will suffer something at the hands of time, just as we are accustomed also to say that time wastes things away, and everything is aged by time, and oblivion comes with the passage of time; but not that knowledge, youth or beauty comes through the elapse of time. For time is by its nature the cause rather of decay."

Even if the determinist is prepared to forgo the explanatoriness of his favoured schema of explanation, he may be unable to secure a determinist world. It may be that the fundamental laws of physics are probabilistic. If they are, they are directed in time (see § 50), and entropy increases with time. It is only a formal possibility that entropy might remain constant. If it did, a Laplacian intelligence could cope with the world around him, although he might find it intellectually unsatisfying. It would give him no indication of the direction of time. We therefore conclude that entropy must either be irrelevant to the direction of time, or else necessarily, and not merely contingently, be connected with it in the way it is.

The increase of biological organization seems to conflict with the Second Law of Thermodynamics. A short survey of a municipal rubbish dump provides the answer. Men achieve their high level of internal orderliness at the cost of strewing disorder throughout their environment; and, more generally, biological organisms, which do indeed become more orderly with the passage of time, are not themselves physical systems. Organisms cannot exist apart from their environment, and the entropy of the two together increases, thus bearing out the Second Law of Thermodynamics, even though that of the organism, considered by itself, decreases.

Organisms embody the principle of homeostasis, and the evolutionary process seems to embody a principle of homeostasis towards greater and more effective homeostasis. If an organism is to maintain a constant state as regards some variables – e.g. a constant temperature of the blood – in the face of fluctuations in the environment, it must take compensating action: and therefore, although in some respects it can achieve a degree of independence of the environment, it does so at the cost of increased sensitivity to the environment – lizards do not sweat on hot humid days. The greater the homeostasis, the greater the sensitivity. The organism comes to have a more complicated system of receptors, and extracts more information from its environment. Only an omniscient God can be impassible, as the theologians say; only if a being possesses perfect information can it foresee and forestall every change, and never be pushed around by events or be the plaything of circumstance. Independence demands knowledge, and man the most independent of all creatures needs to be the most knowing.

The more perfectly the first condition is to be fulfilled – that agents can alter the future – the more there must be homeostatic organisms; for to have goals is a homeostatic condition [12]. Rational agents differ from organisms not in not being homeostatic, but in being able to choose their own goals, instead of having to seek preselected goals that are, so to speak, wired into

[12] I use the word 'goal' in a very wide sense. It is often important to distinguish states of affairs, such as victory in war, from long standing conditions, such as peace. The former are goals strictly so called: the latter only in a wide sense. See further G. Vickers, *The Art of Judgement* (London, 1965), pp. 31–3.

them. An agent, having chosen his goal, brings about its achievement irrespective (within limits) of circumstance, just as an organism maintains its
homeostatic condition irrespective (within limits) of external environment.
An agent differs from other organisms in having a greater degree of independence and a wider range of possible homeostatic conditions, and therefore
also in having a greater degree of complexity and a wider range of information
that could conceivably prove relevant. If biological organisms are, any of them,
to be capable of being agents, the fifth condition follows from the first; and
for those organisms that are, in part at least, rational agents, the second
condition is a special, and crucially important, case of the fifth.

We thus see that the directedness of time is no fortuitous feature but a
fundamental one, connected with the central parts of our whole conceptual
structure. Philosophers have often been misled by the analogy with space. It
is easy to visualize a one-dimensional space, and suppose that time must be
like that. And then we are faced with the intractable problem of how to insert
a direction into this space-like dimension. But this is to put the cart before the
horse. As we shall see, it is not so much that time is like space as that space is
like time. We can obtain a transcendental derivation of space from time. We
shall then see space as time-like, but with the direction taken out. It is very easy
to take a direction out. If aRb gives a directed (that is, asymmetrical) relation
between a and b, then the new relation S, defined by

$$aSb =_{\text{Df}} aRb \lor bRa,$$

will give us a symmetrical, undirected relation. This is what we require of space.
Space is necessary to give us room to be different in, to give things opportunity
to change – and to change back again. These are symmetrical relations, and
so space is undirected, and lacks this quintessential characteristic of time.

§ 9
Cyclic time

Kneale offers, as his prime example of a metaphysical statement that is neither nonsensical nor merely verbal, that temporal order is not cyclical [1]. He is surely right in this. But we find it difficult to say why we are certain that time is not cyclical, or what would go wrong if it were.

Part of the difficulty is to distinguish time from change. Many changes are cyclic. Indeed the processes of nature are characteristically periodic (see § 14, pp. 81–3 and p. 81 n. 3, and § 43, p. 206). And so we are tempted to think that the whole development of the universe is periodic, and to pray ergodically for that blissful consummation "when with the ever circling years, comes round the age of gold". And from this Pythagorean periodicity of process it is natural to argue, with Eudemus, that time too would be cyclic [2].

[1]   W. C. Kneale, "Time and Eternity in Theology", *Proceedings of the Aristotelian Society*, LXI (1960–1), pp. 91–2.
[2]   Eudemus *apud* Simplicius, in *Aristotelis Physicorum Commentaria*, 732, 30 (DK58B34), quoted by G. S. Kirk and J. E. Raven, *The Presocratic Philosophers* (Cambridge, 1957), § 272:

εἰ δέ τις πιστεύσειε τοῖς Πυθαγορείοις, ὥστε πάλιν τὰ αὐτὰ ἀριθμῷ, κἀγὼ μυθολογήσω τὸ ῥαβδίον ἔχων ὑμῖν καθημένοις οὕτω, καὶ τὰ ἄλλα πάντα ὁμοίως ἕξει, καὶ τὸν χρόνον εὔλογόν ἐστι τὸν αὐτὸν εἶναι.

("If one were to believe the Pythagoreans, that numerically the same things happened again,

Pythagorean periodicity is determinist. There is nothing new under the sun: "*κατὰ περιόδους τινὰς τὰ γενόμενά ποτε πάλιν γίνεται, νέον δ᾽ οὐδὲν ἁπλῶς ἐστι*" [3]. We cannot act spontaneously or exercise real freedom of choice, because we are bound to do again whatever it was we, or our qualitatively identical predecessors, did last time. If we reject determinism [4], we must reject Pythagoreanism too. Determinism apart, however, the view that the whole universe undergoes a periodic process is entirely tenable, although none of the arguments for it quite succeed. We can conceive of qualitatively identical but numerically distinct states of affairs recurring after some cosmic great year. But they are numerically distinct. It is change that is perfectly periodic, not time that is cyclic.

If time really were cyclic, there would not be a recurrence of events that were qualitatively identical though numerically distinct, but, rather, the events would be numerically as well as qualitatively identical. It would be the selfsame event – not the same sort of event all over again, but the very same event just once. But with this our whole criterion of identity collapses not only for persons and things [5] but even for events [6]; the very way we formulate the theses – that the same events *re*cur, *re*peat themselves – shows that we need to assume that time is not cyclic in order to state the thesis that it is, and hence that the thesis is itself Strawsonianly incoherent. The fundamental reason is the essential egocentricity of time. Time cannot be analysed simply in terms of change, but is a concomitant of consciousness (see above § 2, § 7, p. 37, and below § 15 and § 52, p. 280). My thinking to myself and my speaking to you must be in time; and conversely, if I am to speak of anything else temporal, I must be able to relate it with my own temporal scheme; so much so that, as we have seen (§ 7, pp. 36–8; see also § 55, pp. 303–4), failure of connected temporal reference is an indication of fiction as opposed to fact.

The proponent of cyclic time is faced with a dilemma. Either he disconnects his thesis from contemporary temporal reality, in which case it is consistent, but self-confessedly untrue; or he admits an egocentric Adam to his Garden of Eden, and with him all manner of self-referential paradox and original inconsistency: either it is fiction, or some self-referential argument will apply. I can be told a cycle of fairy tales, in which Pythagorean periodicities are

---

and that I shall have this pointer in my hand and be telling fantasies to you who will be sitting just as you are now, and everything else will be just as it is now, then it would be reasonable to argue that time too will be the same.")
[3] Porphyrius, *Vitae Pythagorae*, 19 (DK14, 8a), quoted by G. S. Kirk and J. E. Raven, op. cit., § 271: "Events recur periodically; nothing is absolutely new."
[4] See above, § 8, pp. 53–5, and more generally J. R. Lucas, *The Freedom of the Will* (Oxford, 1970).
[5] See further Hans Reichenbach, *Space and Time* (New York, 1957), ch. II, § 21, pp. 141–3.
[6] See also M. F. Cleugh, *Time* (London, 1937), p. 225; and G. J. Whitrow, *The Natural Philosophy of Time* (Edinburgh, 1961), ch. I, § 9, pp. 40–1, and ch. V, § 5, pp. 259–60.

described, and within which nothing occurs to distinguish one cycle from another. I am invited to observe, but forbidden ever to be, any of the characters. The observable overt behaviour in one cycle is indistinguishable from that in another, and within the framework of the story a Leibnizian identity between them is impossible to deny. But it is only a story. I cannot fault it; but I cannot be asked to believe it. For if I am, I can ask "Where do I come in?", and whenever I come, I introduce uniqueness, for I am necessarily unique. There may have been, there may be going to be, another person like me in all observable characteristics, but there cannot be, there necessarily cannot be, another me, another person who is numerically identical with me. In the cardinal – though not the ordinal – sense, I am rightly and necessarily Number One. If an alleged account of cyclic time offers a temporal frame of reference in which I can date my own activities, then I can distinguish one cycle from another. Even if in another cycle there was, or will be, some one qualitatively identical with me, he will not be me unless either I can remember being him or he will be able to remember being me. In the former case there is already, and in the latter case there will be, an adequate criterion for distinguishing the two phases of my experience, and hence everything else too. Moreover, there is something wrong with the claim to be able to tell me all about myself: but if a person claims to be able to locate me in cyclic time, he must be able to do just that, for he must know everything I am going to do, as well as everything I have done, in order to be sure that I shall not introduce something new into the system which would distinguish the next cycle from the last. Such knowledge is too deep for him. He cannot tell me all about myself without exposing himself to refutation at my hands; for, however much he tells me, I can always take into account the further fact that he has told me, and be different from whatever it was he had described. There are logical difficulties in giving a complete and fully specified description of a rational conscious agent [7], which preclude my being fitted into any scheme of cyclic time. But if I cannot feature in it, it remains only an idle speculation, which cannot be, and cannot even be conceived to be, true.

Even if the essential egocentricity of time were questioned, there are difficulties about the order of events in cyclic time. If we take 'before' and 'after' in their usual sense, every event will be both before and after every other event; and it will become impossible to date events, or identify them by reference to their temporal ordering. Although it is possible to reconstruct the concepts from some three-place concepts (such as betweenness), or by invoking some metrical notions, they are cumbersome and implausible representations of the

[7] See J. R. Lucas, "Minds, Machines and Gödel", *Philosophy*, XXXVI (1961), pp. 112–27; reprinted in Kenneth M. Sayre and Frederick J. Crosson (eds.), *The Modelling of Mind* (Notre Dame, 1963), ch. 14, pp. 255–71, and in Alan Ross Anderson (ed.), *Minds and Machines* (Englewood Cliffs, 1964), pp. 43–59. For an opposing view, see Paul Benacerraf, "God, the Devil and Gödel", *The Monist*, LI (1967), pp. 9–39; see further J. R. Lucas, "Satan Stultified", *The Monist*, LII (1968), pp. 145–58, and *The Freedom of the Will* (Oxford, 1970), §§ 21–9.

concepts we actually use. Moreover, even if we could introduce an order into cyclic time, we cannot import a direction. None of the five conditions given in the last section will be satisfied. The third – the time-directedness of causal explanations – is the most hopeful: but if time is to be cyclic, all processes must be periodic; in which case they will be susceptible of a Fourier analysis in terms of trigonometrical functions, and will therefore satisfy a differential equation of the second order, which is, as we shall note (see § 49, p. 253), independent of the direction of time. We can explain the position of the planets by reference to a subsequent state just as well as by reference to a previous one, and in the perpetual periodicity of a Platonic Great Year we should have no reason for preferring one to the other. The fourth condition – that entropy increases – clearly cannot be satisfied. If we are to have cyclic time, entropy must remain constant, not increase. So too, the fifth condition – that biological organization increases – cannot hold over the long run. The first condition fails even on the weaker thesis of strictly periodic change and so *a fortiori* for cyclic time. Only the second condition remains, and that too must fail, at least in principle, for an immortal intelligence. If God can remember the past, He can remember the whole. For cyclic time, unlike serial time, is not divided by the present into two separate regions, the past and future; past and future are connected together and are the same. But if past and future are indistinguishable, so too must be the present. There is no sense of the passage of time, if time is cyclic; no 'now', no moment of present thought or utterance. But if there is no 'now', there is no temporal reference at all. And if there is no passage of time from future to past, there is no sense of time at all. Cyclic time is static time, and static time is no time.

§ 10
The measurement of time

Time ought to be measurable: we are conscious of it in our private experience as a magnitude, a *distentio animi* (§ 2, p. 14, n. 14), and I have argued that public time, although not given to us as a magnitude, must nevertheless be thought of as possessing the order-type θ (§§ 6, 7). We might hope to base our measure of public time on our individual assessments of the passage of private time, even though these are somewhat variable from person to person. After all, we vary a lot in our assessments of colour, yet we can nonetheless establish a reasonably objective science of photometry [1]. The rhythms of talking and walking might be invariable enough to provide us with an intersubjective measure of time. St Augustine thought our sense of quantity of Latin vowels would do [2]. But in a world less strictly trained in the quantities of Latin prosody, where English professors lecturing on political theory cram far more into an hour than Southern senators engaged on a filibuster, we have no calculus available that will make nineteen of one man's words equal a dozen of another's. Nor are the synchronized enthusiasms of the dance floor or the pop group sufficiently stable or widely enough shared for us all to be able to coordinate our activities

[1]   J. W. T. Walsh, *Photometry* (3rd ed. London, 1958), ch. IX, esp. pp. 289–91, 310–11.
[2]   *Confessions*, bk XI, ch. XXVI–XXVIII, xxxiii–xxxviii.

by reference to them. There is a further, and much deeper, difficulty. We cannot so to speak pick up one hour and lay it out against another hour, and compare the two in the way in which we compare lengths or compare colours. And since we cannot compare lengths of time like that, it seems that we cannot measure time at all. This is another example of the danger of the spatial analogy (see § 7, p. 37). Or again, St Augustine feels that we cannot measure intervals which are either future and therefore are not yet, or past and therefore no longer are: "*nisi forte audebit quis dicere metiri posse quod non est? Cum ergo praeterit tempus, sentiri et metiri potest; cum autem praeterierit, quoniam non est, non potest.*" ("Unless perhaps someone will dare to say that what does not exist can be measured? While time is passing, it can be sensed and measured; but when it has passed, since it no longer exists, it cannot.") [3] Or again, "*Non metior futurum, quia nondum est, non metior praesens, quia nullo spatio tenditur, non metior praeteritum, quia iam non est.*" ("I do not measure the future, since it does not yet exist; I do not measure the present, since it has no extension; I do not measure the past, since it no longer exists.") [4] But this again is to assume that time is measured directly.

What we in fact do is to assign measures indirectly. We agree on what certain intervals are by fixing on their end points, the instants the intervals lie between. We cannot directly say what an hour is, but we can reach general agreement on the instant when Merton clock struck ten, and the instant when it strikes eleven, and agree that the interval between is the interval we are talking about.

This enables us to *talk about* intervals intersubjectively even though we may not have intersubjective experience *of* them. We can refer to an interval as being *the* interval between two events that are intersubjective. We then can assign magnitudes to certain of these intervals in accordance with certain principles of assignment. That is, we measure intervals of time by having rules enabling us to pick out pairs of instants, and to say that the interval between one pair is equal to, greater than, twice as great as, the interval between another pair. The measure of time is thus partly a matter of convention, but not completely so. It is partly so, because we are assigning magnitudes to intervals defined in terms of instants, and there could be different ways of assigning magnitudes which we were free to adopt. It is not completely so, because the conventions we choose are subject to rational assessment and criticism, and, in particular, must be internally consistent, coherent, and not too discordant with too many people's subjective experience.

Our system of assigning magnitudes needs to be internally consistent. Else it will be unintelligible to others and useless to ourselves. In particular, the logic of the word 'magnitude' imposes its own requirements. If we have two (non-zero) intervals, especially if we have two adjacent intervals, AB and BC,

[3]   *Confessions*, bk XI, ch. XVI, xxi.
[4]   ibid. ch. XXVI, xxxiii.

the magnitude assigned to the two of them together must be greater than that assigned to either separately. Indeed, we may reasonably go further and stipulate that the magnitude of the sum of two intervals shall be exactly equal to the sum of the magnitudes of them each. As it happens, we can ensure that this condition is fulfilled (see § 14, pp. 78–80), although it will emerge – somewhat counter-intuitively – that with velocities the "addition rule" does not always apply (see § 42).

Our system of assigning magnitudes needs to be coherent. Although we could, without inconsistency, assign measures purely arbitrarily, there would be no point in doing so. We could not share it with other people, and it would not be any use to us. Worse, we would not have time to make all the arbitrary assignments required. There is, so to speak, a great deal of time to be measured and only a little time to do it in. For reasons of economy alone, we should have to adopt some *systematic* way of picking out intervals and assigning magnitudes to them, because only a system can be thought up in a finite interval of time and then applied indefinitely often. Likewise, only a system can be shared with other people, only a system can link up with important regularities among natural phenomena. This does not mean that our system of time measurement must be wholly rational and non-arbitrary. We could measure time by reigns of the sovereign and sittings of Parliament, much as lawyers do for dating statutes. This is sufficiently systematic and sufficiently public to be usable, and in some societies the political climate might be more important than the meteorological seasons. Although it would be difficult to give gardening hints, and a man who had taken a lease for the reign of Queen Victoria would, by our standards, have done much better than another who had taken one for the reign of Edward V, VI, VII or VIII, it would be a natural unit for a court fashion expert to use, or for the Vicar of Bray. All we can say is that our measurement of time cannot be completely unsystematic or arbitrary, and that there is a natural pressure towards having it simple, useful and unified.

Coherence is not the only requirement: there must be some correspondence with people's temporal experience, vague and conflicting though this is. Joshua, needing time for extra play in order to complete the discomfiture of the Amorites, said "Sun, stand thou still upon Gibeon; and thou, Moon, in the valley of Ajalon", and the sun stood still for about a whole day [5]. The story is intelligible. Therefore there are some circumstances in which we should say that the clocks stood still. And so there are some requirements, over and above coherence and consistency, that clocks must satisfy in order to qualify as working – that is, as still measuring time. If the clocks assign a zero measure to an interval which we all agree to have been about "a whole day", and in which we were able to fight a battle and avenge ourselves upon our enemies, we are not prepared to believe the clocks rather than our own experience. In the extreme case, although only in the extreme, we are prepared to reject the

[5] *Joshua*, X, 12, 13.

clocks. A single individual may be wrong – we believe the clocks rather than him; but in the unlikely event of our all agreeing with one another and disagreeing with the clocks, then it is the clocks rather than we who are wrong. For, as we have seen (§ 2, pp. 9–10), time is not what the clocks say, but what they are trying to tell, are there to tell. It is therefore logically possible for the clocks to be wrong, not on grounds of any internal inconsistency or incoherence but for failure to correspond with the common facts of temporal experience [6].

Essentially, then, we measure time by having rules enabling us to pick out pairs of instants, and to say that the interval between one pair is equal to, greater than, twice as great as, the interval between another pair. Indeed, it is sufficient if we could pick out enough pairs of equal intervals that are sufficiently small. But how are we to pick out equal – isochronous – intervals? What grounds have we for assigning the same magnitude to one interval at one time and to another interval at another time? A lot depends on the answer. Our theory of clocks has a great influence on our understanding of time. Once we decide to measure time by external events and observable processes, we are led to think of it as the numerical aspect of process and the dimension of change, and to impose on it – and hence also on space – certain profound and philosophically puzzling properties.

[6]   See further, G. D. Yarnold, *The Moving Image* (London, 1966), ch. I, pp. 14–16.

§ 11
Calendars and clocks

According to the argument of the previous section, a clock is essentially a calendar. It produces a number of publicly observable events – the clock striking or the hands moving – which act as the end points of the intervals to which we want to assign magnitudes. Most people, however, would see the dependence going the other way. Our calendars – i.e. our systems for referring to dates – are based on clocks, rather than our clocks being based on calendars. We use our measures of time – year, month, day, hour – to date things, rather than date them in order to measure them. Both views are correct. The interplay between calendars, which date, and clocks, which measure, is complicated and gives rise to many confusions.

We have shown that temporal instants have the order-type θ of the real numbers, and hence it seems natural to use the real numbers as *labels*, to refer to, or name, *instants* in a systematic and comprehensive way. We also find it natural to use real numbers to describe, or measure, *intervals*. Usually we conflate these two uses and assign that real number as the name of an instant which is also the measure of the interval between that instant and some other conventionally chosen one – the birth of Christ, the flight to Medina, the first Olympiad, or the founding of Rome. We are basing our calendar on a particular clock, rather than defining a particular calendar as our standard clock. It is perfectly satisfactory in practice, but raises a series of puzzles about whether our dating system is relative or absolute. Exactly the same problems arise about space (see below, § 30, pp. 144–5), and so too about orientation, velocity and acceleration. We need therefore to walk very warily, and be very clear exactly what is, and what is not, arbitrary or conventional when we use real numbers *both* as names referring to instants, positions, directions, velocities or accelerations, *and* as measures assigning magnitudes to intervals, distances, angles, relative velocities or relative accelerations. It makes no sense to talk of a point being simply "distant" without being able to say "distant *from what?*"; or to talk of a line being at an angle without being able to say "at an angle *to what?*". 'Distant' and 'at an angle' are relative terms, and can enter into intelligible statements only if both terms between which the relation is alleged to hold are either explicitly given or implied by the context. When, however, we are talking of dates, positions and directions the same considerations do not necessarily apply. It is not part of the meaning of the words 'date', 'position' or 'direction' that they are really dyadic terms which can be properly applied only if two instants, points or lines are specified. We can refer simply to *this* position or *this* direction; or to the position occupied by James or the direction of Jean's gaze, or to the time when they finally get married. We can; and the fact that we can shows the logical difference between instants and intervals, between positions and distances, and between directions and angles. But the difference is obscured when we do not refer to particular positions or directions by means of individualized names or pronouns, but by means of a systematic scheme of reference which relies on intervals, distances and angles, and therefore is implicitly relative. What we are talking about is not relative, but the way we talk about it is. There is an analogy with ordinary language where we seldom refer to particular things by proper names or bare demonstrative pronouns, but usually use some descriptive term to characterize what we are referring to. We talk only seldom of Fido and often of the dog. And hence we are sometimes tempted to infer that particular objects do not really exist, but are really only instances of those universals that are involved in the classificatory system which we feel impelled to use in referring to them. But the conclusion does not follow, and neither does it follow that instants or positions or directions are of necessity only relative. What is true is that our standard ways of referring to them are in some sense relative. We refer to an instant by the measure of the interval between it and some other, generally agreed, base date – the birth of Christ, the flight to Medina, etc. Similarly we refer to the position of a point by giving the distance between it and the axes of some coordinate system we have agreed upon, and to the direction of a line by the angles between it and the respective axes. In each of these cases the way we refer to points is relative to some scheme or frame of reference, and is, to that extent, relative.

This may seem an entirely trivial relativity. After all, every reference system depends on the language being used, and is to that extent relative. But the relativist is saying more than that. He is not saying merely that the meaning of a date – say 1970 – depends on the dating system (1970 A.D. refers to a different date from 1970 A.U.C.), but that any dating system must assign dates by measuring intervals, and these are inherently relative. French is a different language from English, and a Frenchman refers to Fido not as 'that dog' but as '*ce chien*'. But the relativist is saying something more – that whether we speak in French or English we could refer to Fido only in virtue of his canine features and the family resemblance he bore to the whole doggy race. And this, if it were true, would be a substantial restriction on how we could talk. So too, the relativist argues, and with much greater show of reason, that we can refer to a date, a position or a direction only if we can specify the measure of certain intervals, distances or angles. Otherwise, if we do not give a measure, our words are idle.

There is an important interplay between 'I' and 'we'. I can conceive of many things that I cannot know, and in framing my concepts it is what is conceivable that is important rather than what I can be sure I know. But when I come to communicate my thoughts to you, I can do so only if you are in a position to know what I am talking about. It is not enough that I should be able to conceive of something; I must enable you to tell what it is I am talking about; that is, to identify it from my account of it. Hence a common language is much trammelled by what we can know rather than what I can conceive. I can conceive of this instant t or this point (x, y, z) or this direction (ϕ, θ) being different – being t' or (x', y', z') or (ϕ', θ') – but unless I can tell you what difference it makes to some interval, or some distance, or some angle, I am not saying very much; and if I do say what difference it makes, I must say what other instant or point or line the interval or distance or angle is between. And this is what the relativist maintains. And so when the Newtonians conceived of absolute time and absolute space and the universe as a whole being differently disposed in time or space, Leibniz asked "what difference does it make?", and assuming that the answer was "none" concluded that it was an idle fantasy [1].

But the argument is unconvincing. For although the only *systematic* way of referring to instants is by measuring intervals, it is not the only way we have; and although the use of real numbers to refer to intervals is highly appropriate in view of the topology of time (see §§ 6–7), the real-number continuum does not exhaust the properties of time, and time is not adequately characterized by means of the real numbers alone. In particular, we can use the word 'now' as a way of referring to a temporal instant, and it is essential to any scheme of temporal reference that we should be able to identify some date within it as being the date now (see § 2, pp. 7–8, § 7, p. 37). Hence we can obtain enough purchase on Clarke's speculation for it not to be an entirely idle one, nor totally

[1]   Leibniz's Third Paper to Dr Clarke, § 6; reprinted in H. G. Alexander (ed.), *The Leibniz-Clarke Correspondence* (Manchester, 1956).

unintelligible. I can envisage what it would be like for the world to have begun half an hour sooner because I can envisage myself as a conscious being whose "now" was, in our present dating system, half an hour later than it actually is. It is a supposition that I, or any other conscious being, can enter into, although not one I could have evidence for. But since the point of difference is in the use of the token-reflexive, egocentric 'now', the epistemological argument of the previous paragraph does not apply: if Tweedledum and Tweedledee are completely interchanged, I can never tell the difference, and therefore, perhaps, there is none; but this does not mean that if I were you everything would be exactly the same. I know myself in a way different from that in which I know other things; and any concept that is latently egocentric, and in particular the concept 'now', has its application not entirely determined by external discernible criteria and therefore is not subject to the Identity of Indiscernibles (see further below, § 25, pp. 119–20, § 27, pp. 127–8, § 30).

We thus have many different ways of dating, but only one capable of indefinite refinement and extension. We can base our calendar on particular events – James's and Jean's wedding, the accession of Queen Anne, the consulship of Plancus; we can adopt a basically egocentric or token-reflexive style – next new moon, last time I saw you, three years ago, in six months' time; or we can establish a general system which dates all events by the magnitude of the interval between some conventionally agreed origin and the instant at which the event in question occurs. Only the latter is capable of indefinite refinement and extension; but since it is not the only possible one, we do not have to import into our understanding of time the assumptions that such a system presupposes. For, as we shall see, the theory of clocks imposes on time a certain featurelessness which runs counter to our understanding of time as the concomitant of consciousness, the condition of activity or the realization of possibility into actuality.

§ 12
The rational theory of clocks

Although we could use any calendar as a clock, laying it down that every reign, say, was to count as an equal length of time, in fact we have a *rationale* for our system for measuring time, and believe that it is not just our *fiat* which deems periods to be isochronous, but that they really are. Witness to this the fact that our clocks are corrigible. We are prepared to correct our clocks because we have not just laid it down that successive intervals shall be taken as equal, but believe that although we cannot take up a second today and lay it off against a second from yesterday, nevertheless we can have reason to claim that the two intervals have the same measure in spite of their having different dates. Essentially, we adopt some principle of *time-invariance*; we talk about time *timelessly*, so that we can talk about intervals at different times being the same, just as we can talk about colours, lengths or fields at different times being the same. We argue that it will take as long tomorrow for a balance wheel to complete one oscillation as it does today or did yesterday. To quote Sir Edmund Whittaker:

> A first suggestion may be based on an axiom which may be stated thus: any experiment within the domain of science properly so called, may be

repeated as often as is desired, with all its circumstances unchanged except for a displacement in space and time. [1]

or James Clerk Maxwell:

The difference between one event and another does not depend on the mere difference of the times or the places at which they occur, but only on the differences in the nature, configuration, or motion of the bodies concerned. [2]

This gives the rationale of our theory of clocks. In practice, we need to distinguish two cases. Some clocks, the water clock and King Alfred's candles, are linear processes; with them we measure time simply by the volume of water that has flowed or the length of candle consumed. Others (the succession of day and night, of waxing and waning moons, or of the seasons, the oscillations of a balance wheel or of a pendulum, the vibrations of a quartz or of a caesium clock) are essentially periodic, and we measure time by counting the *number* of periods that have been completed. The latter are more economical in the assumptions required to justify them. All we need to assume is Whittaker's axiom. At the beginning of each successive period the initial conditions can often be reasonably supposed to be the same, and the conclusion follows that the interval will be the same too. With the linear processes we need a slightly stronger assumption, that the factor being affected is itself irrelevant to the rate of process – that the amount of water discharged does not affect the rate of flow, that the amount of candle burnt does not affect the rate of burning. In the former case, the assumption is false, and even where it is true it may be difficult to be confident that it is true. We therefore tend to measure time by repetition of periodic processes – giving a metric mesh of order-type $\omega^* + \omega$, rather than by a linear process, in spite of its being naturally of the same order-type (θ) as time itself.

We can combine the advantages of both the discrete and the continuous methods of measurement by using a class of continuous processes that are not linear but periodic. The rotation of the earth on its axis, the orbital motion of the moon round the earth and of the earth round the sun and the rotation of the hands on the dial of a watch are such. The last – in which essentially we measure time by measuring an angle – is the paradigm case. Fourier's theorem shows in effect that every periodic continuous process can be expressed as a function of angles. Where there is such an angle, or something like it, that is a direct function of time, we can measure it as an analogue, of the same order-type as time, while having the further assurance that even if the angle does not change at the same rate throughout the period (as is the case with the earth's motion round the sun), the interval required for each successive complete period will still be the same.

[1]   *From Euclid to Eddington* (2nd ed. New York, 1958), § 18, p. 41.
[2]   *Matter and Motion* (New York, reprinted 1953), ch. I, § 19, p. 13.

We should note, for the sake of completeness, one other mixed case, in the contrary sense: the disintegration of radioactive atoms. This might appear to be a discrete process, since the atoms disintegrate one by one. But in fact we should count it as a linear process, since it is only if we average out the decay over large numbers that we obtain a steady process which can be used as a measure of time.

The principle of time-invariance required for the rational theory of clocks is that of the *date-indifference* of processes. This should be distinguished from the indifference of origin implicit in any metrical calendar. We make an arbitrary choice of origin when we decide to measure A.U.C. or A.D., and either dating system is as adequate as the other. We can translate from A.U.C. into A.D. or vice versa without any essential loss. But of course we shall alter the actual numerals used, as we translate from the one dating system to the other. The date-indifference of processes is something more. It is, as it is intended to be, analogous to the position-indifference and orientation-indifference of a rigid rod. We think that a ruler lays off equal intervals in space, because we think that its length does not depend on where it is: it does not matter where in space one of the end points of the ruler is – the other end point will be the same twelve inches away from it. Similarly, if our clocks are rational, rather than arbitrary, ones, it should not make any difference where we set the zero. It should not matter whether we start our clock now or tomorrow; it should measure off the same interval tomorrow afternoon as being one hour; for then we can say that the hour marked off today by the clock started today would be the same as the hour tomorrow that would have been marked off by the clock had it been started tomorrow, which *is* the same as the hour marked off tomorrow by the clock started today. What corresponds to our being able to move a rigid ruler about in space and lay off a distance anywhere we like is a process's being date-indifferent in time. Not only can we translate – by re-setting the zero – from one dating system into another, but the result of the translation will be to leave the description of the process exactly the same. If I am writing history, I can use either A.U.C. or A.D., but if I switch from the latter to the former, I must increase all my dates by 753: but whether I wind my watch today or tomorrow, and whether I consider the 36,000th oscillation of the balance wheel or the 54,079th, the time taken for one oscillation will be exactly the same. If date-indifference holds, then even though we alter our system for referring to dates, our description of processes will not alter, and therefore neither will the measure to be assigned to the interval taken by a process.

The principle of date-indifference can also be expressed as the principle of the causal irrelevance of time. So far as physics is concerned, we can never explain anything by saying that the time had come for it to happen, or that the time was ripe. Physical explanations are in terms of antecedent conditions, but time itself is not a cause. This, as we shall see, has an important bearing on our concept of time, which, evacuated of all causal efficacy, seems very tenuous and unreal. The tenuousness of time arises from the principle of *date-indifference*, which, although presupposed by the rational theory of clocks, is not given us in experience, nor required by the concept of time by itself, nor even absolutely necessary for time regarded as measured by some arbitrarily imposed metric. We can think of time as a set of instants of order-type θ, each one of which is labelled, the intervals between them assigned by some metric satisfying reasonable conditions of consistency and coherence. The requirement of date-indifference is something extra, which we impose. It is a regulative principle for organizing temporal experience, the first of many principles of symmetry we invoke in making our spatio-temporal schemata as rational as possible. There are good reasons for adopting each of these principles, but their consequences are of great importance, and we need to be very explicit about what we are doing, and why, and subject to what limitations of validity.

One immediate defence of our principle of date-indifference is that it is a precondition of our having any causal laws at all. As we saw earlier (§ 8, p. 53), only if we can have the principle 'Same cause, same effect' can we have causal laws. And 'same cause' must, *inter alia*, mean same at different times. If we are to discover causal laws, we must be able to repeat experiments, or at least observations, that are likely to be the same in all relevant respects. We must therefore have some canons of irrelevance. At the least, we must regard differences of time and personal attitude as being *per se* irrelevant. We could not have causal laws at all, unless we could also have clocks. And unless we could have causal laws, our first and most fundamental criterion for the direction of time would fail.

§ 13
Timelessness, permanence and omnitemporality

Not only causal laws require some principle of time-indifference. It is a necessary condition also for communication and much of thought that a difference of time *per se* should not constitute a relevant difference. Language requires us to concentrate on some differences, which are accounted relevant, and to ignore others, which are to be taken as irrelevant. If nothing was allowed to be irrelevant, nothing could be said. Total truth is ineffable. The most that limited mortals can aspire to is partial truth, selecting some features for comment and passing over others in silence. We must be selective, if we are to talk at all: and therefore some features must be irrelevant. Among them time. For talking takes time. Even in the course of one conversation there emerges some difference of time. If we are to use adjectives at all, they must characterize in such a way that they cannot be held to be inapplicable on different occasions of use. But different occasions of use are necessarily temporally different. Therefore temporal differences by themselves must not be allowed to constitute relevant differences.

This argument from language is one of several that incline us to be Platonists. Although we do not believe in Plato's world of timeless, spaceless and impersonal forms, we do believe that the most important – the most rational –

truths should be expressed by universal statements in the timeless present tense, without reference to any particular places or persons. The paradigm truths are 'two and two make four', 'snow is white', 'acid plus base yields salt plus water'. To say that two and two used to make four in the good old days, or that Fermat's last theorem is true in Cambridge but not in Oxford is, if not a solecism, a joke – like Voltaire's Frenchman who left space a plenum in Paris and found it a vacuum in London [1]. Plato banned all operational terms from mathematics – such as "bisecting" an angle or "squaring" a line – on the grounds that they were not consonant with the timeless nature of mathematics [2]: and the Christian Church took it as a mark of true doctrine that it should be accepted *semper, ubique, et ab omnibus*.

We need to distinguish different ways in which propositions can be independent of time. Some propositions – 'two and two make four' – are timeless in a strong sense: it would be meaningless to raise the question of whether two and two made four in the reign of Queen Anne, or will continue to make four at the turn of the century. In other cases the tenseless present expresses omnitemporality rather than strict timelessness. 'Snow is white' is not a timeless truth about timeless entities, but a universal truth, true everywhere and everywhen, about changeable substances. 'Snow is white' could be contradicted by, say, 'This is snow and is red'. The latter 'is' is a normal tensed 'is', indicating the present tense. It follows that 'Snow is white' (tenseless) implies, among other things 'This snow is white' (tensed present). It also implies 'This snow was white' and 'This snow will be white', since it can be contradicted by the negation of either of these.

In other cases we should speak of permanence rather than omnitemporality. Matter, according to the materialists, is permanent.

> *Nullam rem e nihilo gigni divinitus unquam.* [3]
> *Huc accedit uti quidque in sua corpora rursum*
> *dissolvat natura neque ad nihilum interemat res.* [4]

In saying that matter is permanent, the materialists are saying that it might change, it might come into being or pass out of existence, but it does not [5]. It is perfectly intelligible to speak of its doing so – indeed Hoyle, in his cosmological thesis of continuous creation, supposes that it does come into being [6].

[1] *Letters Concerning the English Nation* (London, 1733), Letter XIV (*Lettres Philosophiques*, ed. Gustave Lawson (Paris, 1964), vol. II, p. 1), quoted by Sir Edmund Whittaker, *From Euclid to Eddington* (2nd ed. New York, 1958), § 6, p. 12.
[2] *Republic*, VII, 527a; see further § 32, p. 151.
[3] Lucretius, *De Rerum Natura*, I, 150: "No thing is ever begotten of nothing by divine will."
[4] ibid. I, 215–16: "From this it follows that nature reduces each thing back again into its elements, but never turns things into nothing."
[5] For a slightly different analysis – the "transtemporal *is*" – see N. Rescher, "The Logic of Chronological Propositions", *Mind*, LXXV (1966), pp. 75–6.
[6] F. Hoyle, *The Nature of the Universe* (Oxford, 1950), and "The Steady-State Universe", *Scientific American*, CXCV (1956), pp. 157–66. For the contrary view see, for example, H. Dingle, "Cosmology and Science", *Scientific American*, CXCV (1956), 234–6.

We can argue for at least a relative permanence of substance on grounds similar to those we argued for the omnitemporality of features. Language would be impossible if substantives could not continue to refer to the same thing throughout a conversation, just as much as if adjectives could not be applied independently of time. We shall develop this argument later (§§ 19–20). At present we need only distinguish permanence from omnitemporality, and both from timelessness in the strict sense. Many confusions, particularly in logic and theology (see further below, §§ 54, 55), have arisen from the assumption that timelessness must be all of one piece.

Of these different sorts of independence of time, it is the second that is important for the theory of clocks. We argue that a natural law, if it is to be rational and part of the framework of reality, must be couched in the impersonal third person, in the omnitemporal tenseless present, and without any particular spatial reference in it. For the first and second persons, the ordinary tenses, and (as we shall see – §§ 30, pp. 144–5) our frames of reference for space are implicitly token-reflexive, and therefore egocentric. But egocentricity is opposed to rationality. What is egocentric depends merely on my whim or arbitrary *fiat*. It is essentially arbitrary, and therefore cannot be rational. What is rational is universal, and is true at all times, in all places, for all persons. And so our ideal for a natural law is that it should not depend on or vary with time or person or space. But, of course, a natural law is normally concerned with changes, which must involve time, and often involve space as well. A natural law must talk about time and space. We therefore want it to talk about them without, in some sense, depending on them. We want it to talk about time timelessly and space independently of location. We do this in two stages. First, we make the dependence on time explicit, and formulate our ideal natural law, correlating the values U_1', U_2', U_3', ... U_n' of a set of n variables of a system at the date t in terms of t and of their values U_1, U_2, U_3, ... U_n at the outset.

$$U_1' = f_1(U_1, U_2, U_3, \ldots U_n; t)$$
$$U_2' = f_2(U_1, U_2, U_3, \ldots U_n; t)$$
$$U_3' = f_3(U_1, U_2, U_3, \ldots U_n; t)$$
$$\cdots\cdots\cdots\cdots\cdots\cdots\cdots\cdots\cdots\cdots\cdots$$
$$U_n' = f_n(U_1, U_2, U_3, \ldots U_n; t)$$

where $f_1(\ )$, $f_2(\ )$, $f_3(\ )$, ... $f_n(\ )$ are a set of $(n+1)$-ary functions. In this way of speaking, the values U_1', U_2', U_3', ... U_n' depend on time, but the expression of their dependence does not. All reference to time has been made explicit, and there is no further implicit dependence on time, and the natural law, once stated in this form, is time-independent. Such a law might of course be false; it might be proved to be false; but it could not *become* false. It is, in a sense, a timeless account of temporal phenomena. We should understand Plato's saying "Time is the moving image of eternity" as meaning that temporal phenomena constitute varying instantiations of eternal, that is to say, omni-temporal, truth.

The formulation of laws of nature in the previous paragraph is only mini-mally independent of time. The functions employed do not change their truth value with time, but only because they make their own dependence on time quite explicit: the value of $U_1', f_1(U_1, U_2, \ldots U_n; t)$ depends on the value of t, which appears explicitly as an argument in $f_1(U_1, U_2, \ldots U_n; t)$. Although we could use this or any other such function to mark off successive intervals which we could deem to be isochronous, such a calibration would be an arbitrary calendar rather than a rational clock. To be rational, we have to say why the intervals should be regarded as isochronous. Nor can we secure the independence of time that we require by seeking natural laws which do not depend on the time variable at all. There are many such laws. But if they are completely independent of time, they will not serve to mark off intervals of time at all. The principle of time-independence we require is more subtle than that. Our clocks must both tell the time, and tell it the same at one time as at another.

We can express date-indifference in a number of different ways. Essentially, we want to say that the final values U_1', U_2', U_3', $\ldots U_n'$ in the natural law formulated above should depend on $U_1, U_2, U_3, \ldots U_n$ and only the *difference* of dates, not the absolute values assigned to them. Mathematically, we could say that the natural law should be formulated in terms of dt, the difference of date, not t itself. That is, U_1' should be given by the equation

$$U_1' = f_1(U_1, U_2, U_3, \ldots U_n; dt)$$

and similarly for U_2', U_3', etc. Granted the continuity of time and the other variables, $U_1, U_2, U_3, \ldots U_n$, we can re-express this as a differential equation:

$$\frac{dU_1}{dt} = f_1'(U_1, U_2, U_3, \ldots U_n), \text{ and similarly}$$

$$\frac{dU_2}{dt} = f_2'(U_1, U_2, U_3, \ldots U_n), \text{etc.} \tag{I}$$

And this represents one fundamental way of expressing date-indifference, namely in terms of differential equations in which t itself (as opposed to dt) does not appear [7].

Alternatively we can express date-indifference in terms of invariance under translation. If we translate the dates from one system to another that differs only in having its origin earlier or later, the equations describing the process should remain the same.

We can express this condition mathematically if we consider the natural law formulated in the previous section as a transformation in n-dimensional

[7]   See Henry Margenau, *The Nature of Physical Reality* (New York, 1950), § 19.5, p. 403.

"phase space" of the n variables, or "coordinates", $U_1, U_2, U_3, \ldots U_n$, into another n coordinates, $U'_1, U'_2, U'_3, \ldots U'_n$. Each point in the n-dimensional phase space represents a conceivable state of the system, which will be transformed after an elapse of time into some other point. Let us call the transformation that will result after an interval a T_a and the transformation that will result after an interval b T_b, etc., and let us call $T_a \times T_b$ the transformation that results from first applying the transformation T_a, and then, on the result of that, the transformation T_b [8]. We then require that

$$T_a \times T_b = T_{a+b}. \qquad (II)$$

For what this condition comes to is that it does not matter how a transformation T_{a+b} is divided up, but only that the total interval should amount to $a + b$, which is tantamount to saying that these transformations are date-indifferent as regards composition, the resultant depending only on the total interval. Or, more formally, we add the magnitude a to all our clock readings: then the transformation $T_b U$ becomes $T_{a+b} U'$, where U' is measured at a zero a earlier than U; i.e. $U = T_a U'$:

So
$$T_b U = T_b (T_a U') = (T_a \times T_b) U'$$
$$= T_{a+b} U'.$$

Hence
$$T_a \times T_b = T_{a+b}.$$

In either formulation the condition seems a reasonable one, which we could hope to use as a criterion of adequacy for a process's being a rational clock, and which we could, with some justification, assume to hold in a good many cases. It is at first somewhat surprising that the condition fails as a criterion because it is too easily, in fact universally, satisfiable. Date-indifference is not a simple regulative principle we impose on our understanding of time, but a complicated tissue of *fiat* and fact, which we need to disentangle in detail.

[8]   Note that some writers express the sequence of T_a followed by T_b as $T_b \times T_a$.

§ 14
Facts and *fiats*

It is not all that surprising, when we come to consider it, that the condition that a natural law be expressible as differential equations should fail as a criterion through being too easily satisfied. If we differentiate sufficiently many times, we can eliminate the explicit variable t from any equation, and reduce it to a differential equation in which t itself does not appear [1]. What is much more surprising is the "regraduation theorem". It turns out that any transformation, depending only on time, of a set of n variables $U_1, U_2, U_3, \ldots U_n$, into $U_1', U_2', U_3', \ldots U_n'$, can be converted into one that satisfies condition (II) on p. 77. For these transformations form a continuous group (see below, § 31, p. 147). And each transformation, T_a, T_b, etc., is defined by the value, a, b, etc., given to only *one* parameter, t. We have, in fact, a group of continuous transformations defined by only one parameter, and can apply a theorem about continuous one-parameter groups, namely that any one-parameter group of transformations (satisfying the usual conditions about continuity and differentiability) can itself be transformed into the "dilatation group", and so is equivalent to a one-parameter group of *translations* [2]. That is to say,

[1]   See, e.g., Henry Margenau, *The Nature of Physical Reality* (New York, 1950), § 19.6, p. 409.

we can find some strictly monotonically increasing function $h(t)$ of t, such that

$$T_{h(a)} \times T_{h(b)} = T_{h(a)+h(b)}.$$

The interpretation of this is that if we have two successive intervals of duration a and b respectively, then we can "regraduate" the measure of time, by means of a regraduating function $h(t)$, so that the effect of transforming first by elapse of $h(a)$ and then by elapse of $h(b)$ is the same as transforming by elapse of $h(a) + h(b)$. The plus sign is our achievement. Clearly $T_a \times T_b$ must be some function of a and b. But the "product" of two transformations $T_a \times T_b$ might well not be given by the simple addition rule T_{a+b}. But, even if not, it can be made to be, by replacing a and b systematically by $h(a)$ and $h(b)$. Thus we can secure by *fiat* the additivity of the magnitudes we assign to intervals (see § 10, pp. 62–3), which guarantees date-indifference, as regards any particular process.

The theorem applies only to one-parameter groups. If our transformations depend on more than parameter – if they depend on the three dimensions of space instead of only the one dimension of time – we cannot be sure of securing simple additivity merely by regraduating our measures. Any manifold that has more than one dimension may have an intrinsic metric of its own that cannot be transformed away by a simple regradation – a fact of considerable importance in relativity theory. But so long as we are considering processes developing in only one dimension we can arrange to measure that dimension in such a way as to secure additivity and origin-indifference. This constitutes both a further argument for the one-dimensionality of time and a source of considerable perplexity for the philosophy of time.

Provided time is one-dimensional we can, it appears, be sure of having a rational theory of clocks. We can be sure, that is, of establishing a metric for time which is date-indifferent, in the sense that the process used to measure time will mark out the same intervals whatever the date at which it is "wound up" and started. *Vis-à-vis* this clock, time is homogeneous. One day is like another, and any time is as good as any other for making a new start, carrying out experiments, setting up initial conditions and seeing what happens. It is an important attribute. We have an apophatic theology of time. We try to view time as being devoid of all features that might distinguish any one time from any other: it is a precondition of our seeing ourselves as totally rational agents, and of our methodology of science. Time is a dimension in which events occur, not a characteristic of events. Only so can events at different dates be qualitatively identical; and therefore only if time has no distinguishing

[2]   W. H. M'Crea, "Note on Group Theory and Kinematical Relativity", *Proceedings of the Royal Irish Academy*, XLV (1938), sect. A, p. 25; and Luther P. Eisenhart, *Continuous Groups of Transformations* (Princeton, 1933), p. 34. For a slightly different approach yielding the same result, see G. J. Whitrow, *The Natural Philosophy of Time* (Edinburgh, 1961), ch. III, § 8, pp. 171–5. I am greatly indebted to Whitrow's discussion of this point. For a simple proof in a special case, see below, § 42, pp. 198–202

character can we apply the maxim "Same cause, same effect" (see § 8, pp. 53–6) which underlies all our activity as agents and all our enquiry as scientists. We need time to be homogeneous, featureless and tenuous if it is to be able to play its pervasive but unobtrusive role in our conceptual structure. We have the same requirements of space, but cannot be sure they will be satisfied. Although we may insist wherever possible in ascribing all causes to factors other than bare spatial position (see below, § 35, pp. 166–8), we cannot guarantee success, and always may find that forces are directly due to the curvature of space, and not to any more extrinsic factor. Time, however, because it is only one-dimensional, can be made always to go straight. And conversely, the need for time always to continue in its *via negativa* constitutes a further argument, if any be required, for the one-dimensionality of time.

But we are puzzled. It is all too easy. If any clock can be regraduated to yield intervals that are date-indifferently isochronous, then even the court calendars of the lawyers and the Vicar of Bray should constitute an acceptable clock, and time might after all be only what is measured by process, with special provisos to take care of the Ajalon case (see above, § 10, p. 63). We feel obscurely uncomfortable when the accuracy of any linear clock is discussed: the water clock, King Alfred's candles, even the earth clock. We feel the force of criticisms – that the candles may gutter, the flow of the water will vary with the head of water remaining in the vessel, the angular velocity of the earth is being slowed down by the friction of the tides. But also we feel that we are being asked to pull ourselves up by our bootstraps, and criticize our clocks by appeal to principles established with their aid. Clearly, we can replace one basic clock by another; but can we meaningfully raise questions about the reliability of what at that day and age is our most reliable clock? Does not the theorem show that all we can do is pay our money and take our choice, and having chosen our fundamental clock, accept it, and graduate our measure of time on the basis of its being a date-indifferent isochronous process?

Part of the answer is an entirely general one. There is no difficulty in a theory of measurement being successively refined by appeal to theories and measurements made with its help. We can take something more or less for granted without taking it to be exactly correct. We could use a water clock to establish the principles of hydrodynamics, and then use these to correct the readings of our water clocks. For the principles depend, and can be made to depend crucially in suitably devised experiments, on other measurements than those of time; and often considerations of theoretical elegance enable us to sharpen our theories of measurement beyond what is given us in experimental fact. Our reasons for believing that the head of water or the friction of the tides is relevant are based on a wide range of considerations which do not depend at all closely on the measurement of time, and can therefore be called in aid to correct our measurements of time. More fundamental is the point that the principle of date-indifference is not the only mark of a function's being the expression of a law of nature. Although we believe that a law of nature can be expressed as a differential equation that does not explicitly depend upon time, not every such differential equation is taken as an expression of a possible law of nature. We are fairly tolerant towards first- and second-order differential equations, but decidedly cagey thereafter, and nothing above a fourth-order one has a serious chance of being accepted. The extreme permissiveness of the principle of date-indifference is held in check by other principles equally important.

More important still is the fact that we can regraduate, in order to secure additivity and date-indifference, only once; and the regraduation required will depend on the transformation in question, which in turn depends on the sort of system it is and the laws of nature involved. We cannot be sure, therefore, that every sort of process will yield equivalent time scales. But our rational theory of clocks has to be a theory of clock*s* in the plural, not one clock in the singular. Although astronomical phenomena have, as a matter of fact, gone far to providing us with one single clock we all can use, our Platonic principle allows that there could be many moving instantiations of omnitemporal laws of nature; and all these should manifest the same uniformities. Ideally, we go a stage further, and hope that not only different instantiations of the same principle but clocks based on very different principles, caesium clocks as well as those based on the conservation of the earth's angular momentum or orbital energy, should all tell the same time – that is, isochronous intervals whose relation to one another was, at all different dates, the same. This, if it be true, is a fortunate fact rather than something that can be secured by *fiat*. It is, to use Whitrow's felicitous expression [3], a *hypothesis*, that *there is a unique basic rhythm of the universe*, and facts could always emerge which would show it to be false. Indeed, some cosmologists have suggested that there is a very slight discrepancy between the *t*-scale and the *τ*-scale [4]. If this were so, then not all laws of nature would be date-indifferent, and there would be some naturally given zero date after all. But it would still be a matter of argument which of the natural laws was to be regarded as date-indifferent, and which was to be taken as suggesting an origin in time. Time in itself is featureless, and can be measured in many different ways, and does not have to be measured in any one particular way. Speculative cosmology apart, however, all our different ways of measuring time agree; and therefore, although it is our *fiat* that makes time homogeneous, it is a fact, a very fortunate fact, but one that could have been otherwise, that it does not matter with respect to what system embodying what natural law we homogenize time, the result is the same. We can *make* any *one* process date-indifferent: and in *fact all* are.

[3]   G. J. Whitrow, *The Natural Philosophy of Time* (Edinburgh, 1961), ch. I, § 10, p. 46.
[4]   Especially E. A. Milne, *Kinematic Relativity* (Oxford, 1948). Also by P. A. M. Dirac, *Proceedings of the Royal Society*, series A, CLXV (1938), 199; E. Teller, *Physical Review*, LXXIII (1948), 801; M. Johnson, *Time and Universe for Scientific Conscience* (Cambridge, 1952); and D. H. Wilkinson, *Philosophical Magazine*, III (1958), 582.

It might seem that a sufficiently drastic use of the regraduation theorem could be relied on to achieve complete homogenization of any number of processes, since any two systems, even if they were instantiations of radically different natural laws, could still be considered as one larger one by a sort of shot-gun marriage, and we could apply the theorem to the transformation for the combined system, and so obtain a homogeneous time scale for the combined system – that is, for both the original systems – and hence for all sorts of systems, all taken together. We can. But the cost could be high. For most of the systems we naturally think of as defining a scale for time are periodic, and periodicity is a topological property in our n-dimensional phase space, which is unaffected by any regraduation we may carry out. If we consider any one system by itself, it is clear that any regraduation will leave the intervals of every period equal: for if T_a transforms $U_1, U_2, \ldots U_n$ into $U_1, U_2, \ldots U_n$ again, so will $T_{2a} = T_a \times T_a$, etc. But the argument will not apply if the system is artificially embedded in a larger one: for although the $U_1, U_2, \ldots U_n$ have been transformed back into themselves, the same need not be true of the further $V_1, V_2, \ldots V_m$ which have been also incorporated in the larger system; nor, after regraduation, can we assume that the transformations of the Us and Vs are independent of each other. Regraduation may involve $U_1', U_2', \ldots U_n'$ being functions not only of $U_1, U_2, \ldots U_n$, but of $V_1, V_2, \ldots V_n$ as well. Hence, even if the U's have completed a cycle, it does not follow that the V's have; and so the interval assigned to the second U-period need not be the same as to the first, in virtue of the facts that the V's are different, and that we have made them relevant. Our normal theory of periodic clocks does not require the regraduation theorem, but does require the effective independence of each "clock" from the changeable features of the rest of the universe (cf. § 12, pp. 69–72). Therefore, although we could use the regraduation theorem to impose on all the systems in the universe one uniform time scale, a Platonic Great Year (which might, however, not be periodic), we might find that we had made each "clock" consistent with every other at the price of having made it, according to our ordinary way of thinking, inconsistent with itself; or, at least, at the price of having destroyed the notion of effective causal independence, which is an essential feature of our scientific thinking.

Independence is a stronger condition than simple date-indifference. We not only lay it down that the date, and if at all possible the spatial location, shall have no influence *per se* on a process, but believe that many other factors are, as a matter of fact, irrelevant. Unless there were a presumption of irrelevance, with regard not only to differences of time, place and personal attitude but to an innumerable range of other factors as well, we could not discover natural laws or causal principles at all [5]. If we are to conduct experiments or carry

[5] See J. R. Lucas, "Causation", in R. J. Butler (ed.), *Analytical Philosophy*, 1st series (Oxford, 1962), p. 47; and E. Meyerson, *La Déduction relativiste* (Paris, 1925), pp. 106–7. See further below, § 35, p. 167.

out practical plans, we must be able to assume that for the most part and in the absence of some reason to the contrary we are dealing with more or less isolated systems; else astrology is a serious science, and the result of every experiment and the fruition of every plan depend as much on the conjunction of the planets and the disposition of the stars in their courses as on the sublunary factors available to our observation and our control. But once we have reason to discount most factors as irrelevant, and to consider not the enormous phase space in which every conceivable factor is relevant but only projections on to phase spaces with a small number of dimensions representing only a few factors, we then often obtain the curve coming back on itself, which gives our argument topological purchase in establishing our metric, and enables us to go much further in establishing a rational theory of clocks.

This is why we prefer periodic processes for measuring time, in spite of the fact that they have the wrong order-type – ω or $\omega^*+\omega$ – rather than θ. Granted independence, we can be more sure that every swing of a pendulum or oscillation of a balance wheel or vibration of a caesium atom takes as long as any other than we can be that any linear process is really uniform. And so we prefer, as it were, digital to analogue computers for computing the passage of time, because the digits have a greater guarantee of uniformity than any analogue could ever have. In fact, as we have seen (§ 12, p. 70), we can obtain the best of both worlds by using trigonometrical functions – those based on angles – which are both continuous and periodic. Indeed, we think it characteristic of time, in contrast to space, that it should be articulated into continuous periods. If I arrange to meet you at the "same time, same place" the "same time" is qualitatively identical with respect to some periodic measure but numerically distinct, whereas the "same place" is numerically identical as well. We use the word 'period' of temporal, but not spatial, intervals. We do not expect periodicity in space, although we sometimes find it (in crystals and wave phenomena), nearly so insistently as we expect it and find it in temporal recurrences. It is partly, although only partly, a further concomitance of independence. Isolated systems are bounded. If there are only a finite number of variables that are bounded, *and* if they can take only discrete values, it follows at once that there are only a finite number of possible states of the system, and that in the course of infinite time, if the system is to continue to change, it must do so by coming round to a previous state and then repeating its course. If the variables are continuous the argument does not hold, although it can be shown that the process must be almost periodic if not perfectly so. Hence we should expect many processes to be almost periodic, and atomic phenomena, where the condition of discreteness does hold, to be perfectly periodic. It is for this reason that we regard atomic clocks, like the caesium clock, as more reliable than astronomical ones, where the periodicity does not have to be perfect, and almost certainly is not.

The rational theory of clocks is thus not reduced to triviality by the re-graduation theorem. The regraduation theorem shows that we always can measure time so as to have one particular clock ticking out equal intervals of time, but, as we shall have occasion to note later (§ 45), it remains an empirical question whether other clocks tell the same time. It is a fortunate fact – if indeed it is a fact – that this is so, and that we can isolate a number of independent clocks all of which in a sense agree. The principle of date-indifference, like that of causality, is neither simply given in experience nor simply imposed on experience; but, the world being what it is, there are good *a priori* reasons for adopting it. If we do adopt it, we are then committed to a number of important consequences: in particular, that so far as physical processes go, there is no reason for preferring any one origin for our dating system rather than any other, and every calendar is of order-type $\omega^*+\omega$ with an arbitrary choice of origin; and that measured time is periodic. These, in conjunction with a further feature we are led to investigate, are what make the time that is the numerical aspect of process and the dimension of change tenuous, but all-pervasive, subject to profound conservation laws but singularly insignificant, and finally so featureless as barely to exist at all.

§ 15
The tenuousness of time

The rational theory of clocks requires that time be date-indifferent, but not that it be continuous. And even if for other reasons we have to think of time as continuous, the periodicity of measured time coupled with the fact that "there is a unique basic rhythm of the universe" makes us wonder whether there might not be a naturally given unit of temporal duration even though there is no date naturally given as the zero – i.e. the origin – of our scheme of temporal reference. But this seems not to be so. Although atomic clocks are all perfectly periodic, their periods are not all the same, nor all multiples of some common basic interval. Although we may be led to change our mind – and would, if the existence of the chronon were established (see § 6, pp. 31–3) – the whole tenor of quantum mechanics is against the hypothesis that all eigenvalues are commensurable, or that there is any basic interval of time. Although all the processes of nature keep time, there is no one master rhythm, beating time for them all, whose period would constitute a naturally given unit of time. We can measure in units of vibrations of a caesium atom, days, months, years, or Olympiads; and having adopted any one of these, we can go further, and multiply (one Olympiad equals four years) or divide (one second equals 1/86,400 of a day), in any way we like. In this, time resembles distance in Euclidean geometry, and contrasts with angle or with distance in spherical geometry (the distance round the world – the circumference of a great circle – is a natural unit in spherical geometry and navigation; we may choose to use others, but there is one which is suggested to us, and which makes all our working specially simple). But with time (and, as we shall see, with space) there is no special unit suggested. Other physical quantities do have naturally given units. Speed (whose dimensions are $L\,T^{-1}$ in the usual notation) has the speed of light as a universal constant, and action (whose dimensions are ML^2T^{-1}) has Planck's constant. But these are not enough to determine them all – a third, independent one would be needed – and time (and space) are in themselves unit-indifferent.

Time should thus be contrasted with simple angles, which have a natural zero and a natural unit, with temperature, which has a natural zero but no natural unit, and with periodic processes, which have a natural unit but no natural zero. It is, thus far, like space or potential energy; not only can we add any constant without its making any difference, but we can multiply by any constant and construe it as only a change in convention. In the language of the mathematicians, our measurement of time is – so far as nature is concerned – determined only up to any linear transformation. This is not to say that time is scale-invariant in the same way as it is origin-invariant (i.e. date-indifferent). Indeed, it cannot be. Although it does not matter when a process begins, it does matter how long an interval has elapsed since it began – else there would be no need for time to enter into the equations of motion at all. Time is not scale-invariant in the way that geometrical properties are scale-invariant in Euclidean space (see below, § 38, p. 184). It might be that time and space together were scale-invariant, although the development of physics does not lend support to the suggestion. But whether or not time is scale-indifferent, it is unit-indifferent. Nature gives us no one unit of time in preference to any other. This leads us to say that time is homogeneous. Homogeneity is a form of mathematical featurelessness, and it is part of our notion of time as the dimension of change that it should be a matrix in which events occur, but not having in itself any bearing on the course of events, and therefore as featureless as possible (compare below, § 35, pp. 166–7). Indeed, its featurelessness could be said to be one of its chief peculiarities [1], and has led some – notably Newton – to argue that it must for this reason be something absolute, over and above any events that occur *in* it, and others – like Leibniz – to deny that it is anything real at all. The homogeneity of time is not simply a fact forced on us by experience, but, like continuity and date-indifference, also a *fiat*, something we impose on our schemata for organizing experience. We view things in a certain way in order to accord with the topology required of it as the concomitant of consciousness, and to secure the sameness we require for our rational theory of clocks; and then discover the same uniformity and sameness

[1] See, generally, Millič Čapek, *The Philosophical Impact of Contemporary Physics* (Princeton, 1961), ch. III.

in things, and either announce it as a conservation law or rename it tedium, and either boast that we have revealed the underlying uniformities behind the flux of phenomena or complain that time is indifferent to our affairs, not realizing that in both cases we are discerning in our experience of time only what we had ourselves constructed into it for our own special purposes, and that we should neither be surprised if we detect it to be regular nor think we have anyone to blame but ourselves for an indifference which we insisted on.

Date-indifference yields conservation laws. We establish first the slightly surprising conclusion that not all states are physically possible for any system whose initial conditions are given. That is, although for any set of values of U_1, U_2, ... U_n, the group of transformations T_t will yield other, indeed non-denumerably many, sets of values, it will not yield them all. For consider the $(n + 1)$-dimensional phase-and-time space of U_1, U_2, ... U_n; t. The course of development from one particular set of initial conditions will be represented by a continuous curve – a one-dimensional "space" – in this $(n + 1)$-dimensional space, going through the point representing the initial conditions at the outset. Its projection on n-dimensional phase space is therefore also a continuous curve, which therefore cannot "cover" the whole of the n-dimensional phase space, maintaining a one–one correspondence with every point in it, because there cannot be a one–one correspondence that preserves continuity between spaces of different dimensions (see below, § 29). Of course, as far as the topology is concerned, the one-dimensional curve could be in some many–one correspondence with the n-dimensional phase space. Like Peano's curve (see below, § 29, p. 137) it might go through every point once, but some more than once. But this is ruled out by date-indifference. If ever it comes back to the same point a second time, it must thereafter run through the same cycle over and over again. It therefore cannot be like Peano's curve. It cannot be in a many–one correspondence nor in a one–one correspondence; therefore it cannot "cover" phase space completely. It follows that the group of transformations T_t will partition phase space into a number (perhaps an infinite number, perhaps a non-denumerably infinite number) of equivalence classes (see below, § 28), all the members of each one of which have some characteristic in common. This characteristic depends on the values of U_1, U_2, ... U_n. It is therefore a function of U_1, U_2, ... U_n; and we can argue further for its being a continuous function, whose value is a scalar magnitude. It is clear, however, that if we subjected such a function to multiplication by any arbitrary constant the conservation argument would go through unaffected; equally, if we added to it any arbitrary constant. Hence, what is conserved is defined only up to addition and multiplication. Energy has the same metrical properties as time. Indeed, if energy had not already been discovered, we may say with Voltaire that it would have been necessary to invent it [2].

[2]   See further L. D. Landau and E. M. Lifschitz, *Mechanics* (2nd ed. Oxford, 1969), esp. ch. I and II; R. P. Feynman, R. B. Leighton and M. Sands, *The Feynman Lectures on Physics*

Date-indifference also yields a sense of temporal pointlessness. We do not seem to be getting anywhere if all temporal locations are alike.

> Vanity of vanities, says the Preacher; all is vanity
> What does man gain by all the toil at which he toils under the sun?
> A generation goes and a generation comes,
> But the earth remains for ever.
> The sun rises and the sun goes down, and hastens to the place where it rises.
> The wind blows to the south, and goes round to the north;
>> round and round goes the wind, and on its circuits the wind returns.
> All streams run to the sea, but the sea is not full; to the place where the streams flow, there they flow again.
> All things are full of weariness; a man cannot utter it, the eye is not satisfied with seeing, nor the ear with hearing.
> What has been is what will be, and what has been done is what will be done; and there is nothing new under the sun.
> Is there a thing of which it is said, "See, this is new"?
> It has been already, in the ages before us.
> There is no remembrance of former things, nor will there be any remembrance of later things yet to happen among those who come after. [3]

Moreover, if it does not matter how we set the zero of our dating system, and we might as well reckon our years from the French Revolution as from the birth of Christ, we seem to be denying our belief that if time is to be of any moment to us, some moments must be momentous.

> Then came, at a predetermined moment, a
>> moment in time and of time,
> A moment not out of time, but in time, in
>> what we call history: transecting,
>> bisecting the world of time, a moment in
>> time but not like a moment of time,
> A moment in time but time was made through
>> that moment: for without the meaning there
>> is no time, and that moment of time gave
>> the meaning. [4]

(Reading, Mass., 1966), vol. I, §§ 11, 15, 16, 17, 52. More generally, see E. P. Wigner, "The Role of Invariance Principles in Natural Philosophy", *Proceedings of the XXIXth International School of Physics "Enrico Fermi", Vasena, 1963* (New York, 1964); see also E. P. Wigner, "Symmetries and Conservation Laws", *Proceedings of the National Academy of Sciences*, LI (1964), pp. 956–65; reprinted in *Physics Today* (March, 1964), pp. 34–40.
[3]   Ecclesiastes, I, 2–11, R.S.V.
[4]   T. S. Eliot, *The Rock*, Chorus VII, in *Collected Poems* 1909–1962 (London, Faber & Faber, 1963; New York, Harcourt Brace Jovanovich), p. 177.

An entirely featureless time is emptied of all significance, and the only sort of time that can be fulfilled – Πεπλήρωται ὁ καιρός [5] – is one that allows special times and seasons; and equally it would be as absurd an endeavour to be redeeming some homogenized time – τὸν καιρὸν ἐξαγοραζόμενοι [6] – as it would be to buy back particular items of cash. Theologians sometimes draw a contrast in the Septuagint and New Testament between χρόνος (measured time, duration) and καιρός (time of opportunity and fulfilment) [7]. The linguistic argument is open to attack [8]: χρόνος and καιρός are often used interchangeably [9]. But the conceptual point remains valid, not only for theology but for all humane disciplines. Human affairs are not date-indifferent, but essentially the reverse. For, first, each man at any particular date must regard that date as different in as much as it is present whereas all the others are past or future. We cannot entirely eliminate egocentricity, and the sense that date-indifference does this is one reason for the disquiet it engenders [10]: date-indifferent time must be time that is indifferent to me with my present problems, future aspirations and past achievements. Moreover, it seems unmindful of memory. In human affairs we collectively remember some of the past, and a knowledge of history effectively differentiates dates. For these two reasons, the principle of date-indifference runs counter to our understanding of time when we are concerned with ourselves in particular or human affairs in general, rather than with natural phenomena. It is important therefore to re-emphasize that the homogeneity of time is only a stipulation we have imposed on the way we shall measure public time, not a fundamental feature inherent in the concept of time itself. There are good reasons for adopting methods of measurement that are origin-invariant and impose on time as few mathematical peculiarities as possible; but that does not mean that time in itself is pointless or empty of significance.

The featurelessness of time makes it seem unreal as well as meaningless. Although we refer to it by a noun, often dignifying it with a capital letter, it must be insubstantial if we cannot ascribe to it any properties in its own right.

[5]   Mark, I, 15.
[6]   Colossians, IV, 5.
[7]   J. Marsh in A. R. Richardson, *Theological Word Book of the Bible* (London, 1950), p. 258. See also N. B. Cullman, *Christ and Time* (London, 1951 and 1962), p. 39; and J. A. T. Robinson, *In the End, God* ... (London, 1950), pp. 45 ff., and John E. Smith, "Time, Times and the 'Right Time': *Chronos* and *Kairos*", *The Monist*, LIII (1969), 1–3.
[8]   Especially by J. Barr, *The Semantics of Biblical Language* (London, 1961), and *Biblical Words for Time* (London, 1962); see further G. D. Yarnold, *The Moving Image* (London, 1966), ch. II, esp. pp. 29–36; and F. H. Brabant, *Time and Eternity in Christian Thought* (London, 1937), pp. 235–62.
[9]   At least by St Peter; see his speech in Acts III, 19 and 21, and his letter, I Peter, I, 5 and 20. St Paul writes τὸ πλήρωμα τοῦ χρόνου in Galatians IV, 4, and the author of St Jude ἐπ᾽ ἐσχάτου χρόνου (v. 18). Plato puts the words χρόνος ἦλθεν εἱμαρμένος ("the allotted time came") into the mouth of Protagoras (*Protagoras*, 320d), which he certainly would not have done if there was the slightest sense of solecism in this use of the word χρόνος in classical Greek.
[10]   See, very fully, Richard M. Gale, *The Language of Time* (London, 1968), part II.

Aristotle is inclined to suspect, although for other reasons, "ὅτι μὲν οὖν ἢ ὅλως οὐκ ἔστιν ἢ μόλις καὶ ἀμυδρῶς" [11], and defines time in terms of process (or change). Leibniz concluded "that time, without things, is nothing else but a mere ideal possibility" [12], not allowing that "time was anything distinct from things existing in time" [13]. Time by itself seems too tenuous to talk about, and it is common form to condemn Newton for being metaphysical, in the pejorative sense, in insisting upon an "absolute, true and mathematical time" which "of itself, and from its own nature, flows equably without relation to its own nature" in contrast to "relative, apparent and common time" which "is some sensible and external (whether accurate or unequable) measure of duration by means of motion" [14]. Newton, it is said, has been refuted by relativity; his metaphysics will not stand up to any modern philosophical critique, and anyhow is out of fashion in the age of the common man; and we are told that we should conform to the scientific spirit rather than be left attempting to grasp some thin ethereal entity whose essence is always eluding us.

Nevertheless, the argument against absolute time is not conclusive. The general tenets of positivism are invalid, as we have seen (§ 2, pp. 9–11, § 5, p. 28). The relativity that Newton here rejected is not the relativity that Einstein propounded; and although the Special Theory of Relativity has shown Newton to be wrong in some respects, and in particular has shown that we should not think of time by itself in complete independence of everything external – time is related to space, and also to velocity, contrary to Newton's opinion – it has not shown that time is relative in Newton's sense, and merely some numerical measure of process. More fundamentally, the very featurelesssness of time that makes us suspicious of its existence should be seen as an argument in its favour. Process is far from featureless. The fact that we are prepared to regraduate or to look for further causes in order to ensure that time is homogeneous and that a difference of date cannot, *per se*, explain an event, bears witness to the allegiance the category of time extorts from us. Even if not observed as a phenomenon, it is evidenced in our *fiats*. Our *fiats* are not entirely arbitrary. Although we could establish a purely conventional time scale with any continuous process or discrete succession of changes regarded as our clock (see above, § 10, pp. 62–3), we have in fact a rational theory of clocks, and regard some processes as regular, and therefore suitable to use for measuring time, and others as irregular and unsuitable. And the fact that we have a rational theory of clocks vindicates Newton's doctrine of absolute time. If we really regarded time simply as the measure of process, we should have no warrant for

[11] *Physics*, IV, 10, 217b 33–4: "that either it does not exist at all, or only barely and obscurely."
[12] Fifth letter to Clarke, § 55; reprinted in H. G. Alexander, *The Leibniz-Clarke Correspondence* (Manchester, 1956). See also § 47, and *New Essays*, II, 13, 17, and elsewhere.
[13] Third letter to Clarke, § 6.
[14] *Principia*, Scholium to Definition VIII, 1; reprinted in Alexander, *The Leibniz-Clarke Correspondence*, p. 152.

regarding some processes as regular and others as irregular. All would be equally good. Any continuous process could be made the standard by which we measured off isochronous intervals. It would be simply a matter of convention which one we adopted, not a matter of argument. But we do not regard it simply as a question of convenience whether the solar day, the mean solar day, or the mean sidereal day is really an isochronous interval or not. We are prepared to correct even the mean sidereal day, on purely theoretical grounds and not for any considerations of practical convenience, to take into account the retardation of the tides on the earth's angular velocity; and the caesium clock is in principle subject to the same judgement – if, for instance, calculations in the General Theory of Relativity showed that frequencies were lower in gravitational fields. Even our best clocks are subject to correction. So long as we are prepared to assess the time-keeping qualities of a clock, and are prepared in principle to replace it by a more regular one, if it could be obtained, we are committed to an ideal of absolute time which is not simply what the clocks actually say [15].

Our concept of time is elaborately constructed, as much by stipulation as from experience, and often is made to serve purposes that are not altogether compatible. The time required by the rational theory of clocks has certain desirable topological properties, but is very tenuous, and is emptied of all significance and almost all reality. It does not do justice to the essential ego-centricity – or, better, mind-connectedness – of time, nor to its directedness, nor to its potential purposefulness. And to these topics we shall return when we consider time not only as the concomitant of consciousness and the numerical aspect of process, but as the condition of agency and the transition from aspiration to achievement. But before doing that we need to develop tenuous time further, and show how we can draw out of it the concept, and many of the properties, of space.

[15]   Plotinus (*Enneads*, III, 7, 9, esp. ll. 31–5) used essentially this argument, among others, in his criticism of Aristotle's doctrine that time was the measure of change. Newton would have known of it through Henry More.

II The argument from time to space

§ 16
Space

Space is a difficult concept. It is less mysterious, less mystical, than time, but more intricate, more mathematical. We feel inadequate when we hear physicists saying that space is not flat, but everywhere possesses some definite curvature. We have learned to sneer at Newton for believing in absolute space, at Kant for holding that it must be Euclidean. Among the *cognoscenti* the word 'space' trips off the tongue with the greatest of ease – phase space, sample space, linear spaces, Banach spaces – the world seems full of spaces. But on other occasions we still feel we want to know what space is. Are all things located in space, or only some? Could it be the case that none were? When we describe ourselves as "dwellers all in time and space" do we dwell in each in the same way? Is space as necessary a concomitant to consciousness as time is?

These problems are made more intractable by the fact that the word 'space' can be used in four different ways. It can be used, first, as a term of pure mathematics, as when mathematicians talk of an "*n*-dimensional vector space", a "three-dimensional projective space" or a "two-dimensional Riemannian space". In this sense the word 'space' means the totality of all the abstract entities – the "points" – implicitly defined by the axioms. There is no doubt that there exist, in this sense, non-Euclidean spaces, because all that is claimed by such an assertion is that sets of non-Euclidean axioms constitute possible implicit definitions of abstract entities – that is to say that some sets of non-Euclidean axioms are consistent. If Kant, or any other philosopher, had denied this, he would have been wrong; but there is little reason to suppose that any philosopher, concerned about space, has been using the word in this, the pure mathematician's sense.

The second use of the word is that of the physicist. The word 'line', for example, may be taken to be exemplified by the path of a light ray, or the path of a freely moving particle, or a geodesic (the shortest distance between two points). Under such interpretation, the axioms of some of the theorems will state synthetic propositions which can be put to an empirical test, and it becomes an empirical question whether a particular set of geometrical axioms under a particular interpretation is true or not (see further below, § 33) – and, as is fairly well known, if we interpret straight lines as being the paths of light rays, space turns out to be not Euclidean but Riemannian.

These two are technical senses. In the mouths of the expert mathematician or physicist they should cause no confusion; but often they do give rise to difficulty because we use the same word in ordinary parlance. "This space is occupied" we say to the intrusive passenger on a train, and this gives us breathing space until the diner returns from the restaurant car. Later, when we come to drive ourselves, we are always looking for parking space, and hotels and boardinghouses advertise it among their attractions. Diplomats deal with air space, and the German people, according to Hitler, needed living space. From this, third, group of everyday senses, we move, on reflection, to philosophic ones. Space, we may conclude, is necessary to give us room to be different in: space provides a bare minimum for the possibility of change. Or, again, space is where things can be. Each place is where a thing could be, though it need not be. But everything must be somewhere or other in space, though it need not be at any particular place (see § 29). Space, we feel, is the *barest* possibility of existence, and from this we may argue that it must have the *fewest possible* properties, and be an undifferentiated I-know-not-what (see § 35). But is space anything then? Is it not so devoid of properties that it is entirely insubstantial? So unactualized a potentiality that it is a perfect non-entity? It depends on what conceptual role space is to play. So long as we are merely talking about it, all that is requisite is that we should understand what we say: as soon, however, as we want to use the concept in practical contexts, we shall have to make use of some actual, and not merely potential features, and to that extent compromise our ideal of complete potentiality, altogether unactualized. Verification arguments begin to apply, and we find we cannot say anything true about space unless we have some actual feature, and some way of telling whether what we say is true or false.

Philosophers have varied in the role that they have ascribed to space, and in what they have been prepared to say about it. Some – Poincaré [1], and on occasion, Kant [2] – have claimed that it is a necessary condition, perhaps a subjective condition depending on the nature of the human mind, that the space we actually think of should be a Euclidean space. Space in this sense is neither the pure mathematician's construct nor the remote object of the sophisticated physicist's discovery, but the space of our ordinary experience given to us, and given to us under a standard interpretation, in all our visual and tactile experience. Some connection is claimed between some central aspects of experience and the Euclidean nature of the space in which this experience is given us, that makes it unthinkable that space in this sense should not be Euclidean.

Often, however, when Kant and other philosophers talk about space, they are using the word in a more abstract way, in which it makes no sense to ask whether it is Euclidean or not. Space is a necessary concomitant of other fundamental categories – of time, of substance, of change, of motion. Such a space is more primitive than the one about which it can be asked whether it is Euclidean or not. It still, as we shall show (§§ 26, 28, 29), needs to have certain topological properties – continuity, connectedness, dimensionality – but not any metrical properties, and therefore neither entails nor is inconsistent with there being spaces, in another sense, that are non-Euclidean. Nevertheless, Kant is not guilty of a gross confusion in using the word 'space' both in this abstract sense and in a metrical one, for similar, although weaker, arguments can be adduced for our preferring to have space Euclidean as for ascribing to it certain topological properties. We shall show first how the abstract category of space is connected with other categories, and offer a "transcendental derivation" of philosopher's space from time; that is, we shall show that if we have the concept of time we must also have a concept of philosopher's space. We shall then derive further topological properties of philosopher's space, and establish some surprisingly strong results. We shall then introduce some metrical concepts, and give a justification for the pre-eminence of Euclidean geometry. Since this runs very much counter to contemporary ortho-doxy, some discussion will be necessary of the first and second senses to estab-lish how much, and how little, contemporary views are being controverted.

[1] H. Poincaré, *La Science et l'Hypothèse* (Paris, 1902), ch. III, IV, V, pp. 49–109, esp. pp. 66–7, 79–80, 83; and *La Valeur de la Science* (Paris, 1914), ch. III, pp. 59–95.
[2] See, e.g., H. J. Paton, *Kant's Metaphysic of Experience* (London, 1936), vol. I, ch. VII, §§ 5 and 8, pp. 155–7, 160–3. But see Kant's *Gedanken von der wahren Schätzung der leben-digen Kräfte*, § 10 (I, 24). For general discussion see G. Martin, *Kant's Metaphysics and Theory of Science*, ch. I, esp. p. 18; and also L. Nelson, *Bemerkungen über die nichteuklidische Geometrie und den Ursprung der geometrischen Gewissheit*, Abh. d. Friesschen Schule N.F.I. (1906). W. Meinecke, "Die Bedeutung der nichteuklidischen Geomtrie", *Kantstudien*, XI (1906). P. Natorp, *Die logischen Grundlagen der exacten Wissenschaften* (2nd ed. Leipzig, 1921), pp. 309 f. O. Becker, "Mathematische Existenz", *Jahrbuch f. Philosophie und phäno-menologische Forschung*, VIII (1927), p. 737.

We shall attempt to show the rationale of some other propositions about space and time, commonly believed today to be purely empirical, and in particular we shall consider some explanation for the three-dimensionality of space.

§ 17
Outline of the argument from time to space

The concept of time has often been thought to be like that of space. In many ways, the spatial metaphor is misleading, but it is not totally misconceived. For as I shall now attempt to show, if we have any concept of time then we must have some concept of space. Hence the merits of the metaphor. Hence also its misleadingness. For the argument goes from time to space, not from space to time. We develop our concept of space by analogy with time, not vice versa. Space is like time, but time is, in important ways, unlike space.

That time and space are unlike will be apparent if we consider the two pairs of contrasting statements:

A: (i) A thing cannot be in two places at one time.

   (ii) A thing can be in one place at two times.
and

B: (i) Two things cannot be in the same place at the same time.

   (ii) Two things can be in the same place at different times.

Thus, at least in relation to things, time and space are importantly different. It is through their relationship with things that I shall show how time and space are connected, in what respects they are similar and dissimilar, and why the four statements above are true.

The argument is a complicated one. The diagram below, looking like the formula of a benzene-derivative, indicates various paths of argument. We argue from the concept of time to the concept of two things being qualitatively identical but numerically distinct. We can do this either via the concepts of change and thinginess (the left hand side of the benzene ring) or via that of communication and its concomitant, the type-token distinction (the right hand side of the benzene ring). We can also, if the concept of a thing is felt to be insufficiently

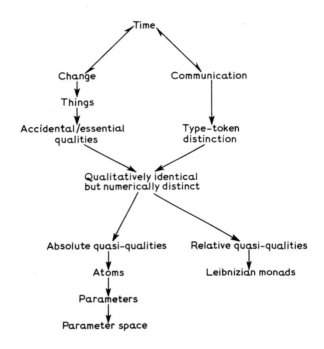

secure, cross over and show that if we can communicate at all we must be able to develop some sort of concept of a thing. Having established the concept of two things being qualitatively identical but numerically distinct, we are faced with two options. One, that taken by Leibniz, leads us to the concept of monads as the fundamental things; the other, that taken by Newton and Locke, leads us to the concept of atoms, as absolute things, and the void, as absolute space. It is thus that we develop one philosophically important facet of our concept of space. We are led to space essentially as the correlative of things. The order of argument is time–things–space. It should be contrasted with the seventeenth century approach of Isaac Barrow, for whom the order was things–time–space [1], and with that of much of modern physics, which starts with space-time and introduces things, if at all, as merely a special sort of field (see below, § 41).

[1] See Isaac Barrow, *Lectiones Geometricae* (London, 1670), *Lectio* I, p. 2; reprinted in *The Mathematical Works of Isaac Barrow, D.D.*, ed. W. Whewell (Cambridge, 1860), II, 160: "*Abstracte loquendo, tempus est perseverantia rei cujusque in suo esse.*" ("Time is, to speak abstractly, a continuance of each thing in its own *being*.")

§ 18
Time, change and communication

Time is a prerequisite of change; also of communication. Many have argued that change is a prerequisite of time, and that if there was no change there would be no time either. But this is a dubious argument: partly, because, as we have seen, we construe time as abstractly as possible, and separate it so far as we can, from the changes that take place in it (§ 12, p. 72; § 13, pp. 75–6; § 14, pp. 82–3; § 15, pp. 90–1); partly because one can conceive of the whole universe being stilled for an interval while one observed it, and then resuming its motions (§ 2, pp. 8–13; § 10, p. 63). One could not measure the interval, but one could be conscious of it, and that seems enough to make the concept of time without change intelligible. But the condition of its intelligibility is that there should be a conscious being to note the passage of time without change occurring; and such a being is a potential communicator. Therefore time without change is possible only if communication is possible. Therefore even if time does not imply change, it does imply either change or at least the possibility of communication. Hence, although we cannot argue either from time to change with certainty, or from time to the possibility of communication with certainty, we can argue with certainty from time to either change or the possibility of communication. One or the other paths of argument must be open to us.

§ 19
Things

The world in which we live is made of things. Things form the context in which our lives are set, they are the material with which we have to deal, that which is both resistant to our will and which alone can carry its impress, that about which we can talk and by means of which we can interact with others. Things are thus centrally important to our thought, and for that reason are difficult to think about clearly. Philosophers have attempted to elucidate the concept under the title of 'substance', but I shall avoid the term. 'Substance', like 'essence', 'quality', 'property' and 'accident', has undergone a sea change of meaning. Under the influence of the chemists it has been transmuted into something much less elemental than Aristotle's πρώτη οὐσία, much less subtle than the quiddity of the Schoolmen. Indeed, we have doubts now whether substance in any of its traditional senses really exists. And the onetime man of substance is now more securely termed a man of property.

The concept of thinginess remains. It has many strands, connected but not always necessarily so. In one sense, anything that can be thought of or referred to is a thing; but in another, many things that can be thought of are not things – not numbers or thoughts or qualities or values, but only chairs, trees, motor cars and the like. Much philosophical perplexity has been generated by this double sense of the word 'thing' – much philosophy has been devoted to proving, or to controverting, that things (in the latter sense) are not the only things (in the former sense) to exist. More generally, things are thought of not only as being the objects of our thought, but as external to us, having an existence of their own, as being resistant to our will and sometimes opaque to our understanding, but as being intersubjectively identifiable, and sometimes rational as well.

Things are external to us, they are what is other than ourselves, they are reality. We long to know the nature of things, and have, on occasion, to come to terms with things. For being other than ourselves, they are resistant to our will; they are characteristically not as we would have them be. Fings ain't what they used to be, the old lag laments. If he had his way they would be different from what they are now; but he is not getting his own way, and they constitute not only what he does not want but a brute fact he has nevertheless got to accept.

For this reason too, not only what is resistant to our will but what is opaque to our understanding seems thing-like. When we understand why something should be so, we enter into the rationale of it, and cease to regard it as an externally imposed necessity because we endorse it thereafter ourselves. Analytic necessities do not necessitate, and even with physical necessities, the more we understand their rationale, the less we feel restricted by them. When I think it through I find I cannot really want two and two not to make four, or for an omnipotent God to be so powerful that he can make a world too big for him to be able to move. And similarly, though to a lesser extent, the scientist does not want hydrogen to be bivalent or the Second Law of Thermodynamics to be false. The knowledge of necessity does not, as some philosophers have claimed, constitute freedom, but the understanding of necessity often alleviates the sense of slavery to external circumstance.

Things are thus thought of as unintelligible, matter as incomprehensible. But on another line of thought, rationality is regarded as a mark of objectivity. Space and time have often been said to be substances by philosophers, because they are preconditions of all thought, or at least all communication, and cannot be imagined away by anyone who would share his thoughts with others. Invariance is a mark of reality. Things are the same for everybody, and are most thing-like when they can be understood to be the same everywhere and for ever. We may have qualms about terming space or time or natural laws substances, but we acknowledge their substantiality by using substantives to talk about them, and often, teutonically, using capital letters as a further sign of ontological respect.

Things are what we talk about. Servants talked about people, but the gentry were supposed to converse about things. Hence their recalcitrance, hence also their rationality. I cannot choose the topic of conversation entirely to suit myself. If I am listening to you, I have to follow your choice, perhaps entirely opaque to me, if I am to follow your discourse; but even if I lead off, I cannot have sole say and be entirely arbitrary in my thoughts, or you will not be able to figure out what on earth I am talking about. I may just manage the Golden Mountain in academic circles, but elsewhere will be disregarded as a square unless I light upon more easily identifiable topics of discourse. It is necessary, as every non-bore knows, to fashion one's speech to the needs of one's listener as well as to one's own interests, if a worthwhile conversation is to be had; and therefore the objects we refer to cannot be the creatures of the speaker's, any more than of the hearer's, will. They cannot be completely subjective: they must be intersubjective, at least to the extent of being intersubjectively identifiable, and to that extent non-egocentric and rational. But beyond that they must be subject to the choice of whoever is doing the talking, and opaque to the will of others and to reason.

The things that I talk about must be not only identifiable, but re-identifiable. For any conversation takes time. If you are to take up anything I say, or if I am to revert to anything I have said, it is necessary for us to be able to refer again to the same thing as we were talking about. To be a thing thus implies the possibility of being the same thing. "Everything is what it is", said Bishop Butler, "and not another thing". Each thing must be capable of being referred to uniquely on more than one occasion, and therefore some things at least must continue through time, and be not only referred to but identified in experience on different occasions. Hence permanence is a paradigm property of thinginess.

§ 20
The argument from the possibility of communication to things

Although the concept of a thing is central to our conceptual structure as it is, some philosophers argue that it is not essential. Instead of having a subject-predicate language in which things are referred to by substantives or substantial phrases, we might make do with "property-location" or, better, "feature-placing" languages [1]. We might have a language containing only impersonal verbs, 'It is snowing', 'It thunders', or 'There is water here'. The last of these examples does offer a viable alternative to thing-language: we can set about describing each part of the universe instead of each particle in it, and this will lead us to modern field theory instead of atomism, and to the plenum of Leibniz instead of the corpuscles of Newton. But in doing this and making places replace things, we find we reify places and make space into a thing. We shall have formidable difficulties, which Leibniz himself foresaw, in identifying things, but if we are prepared to surmount them, we can; but then we find we have not so much eliminated things, as changed the aspect under which we consider them.

The examples 'It is snowing', 'It thunders', do not involve us necessarily in referring to places, but only to temporal dates. It might be claimed that such a

[1]   P. F. Strawson, *Individuals* (London, 1959), pp. 203 ff.

language really was free of all referring terms, since the very context of utterance would be enough to indicate the incidence of the climatic feature characterized by the impersonal verbs used. However, not only would our conversation be very boring if it was thus limited to discussing the weather, but it would be defective in a deeper way [2]. If we could use these verbs only in the present tense, we should not really be talking, but merely producing verbal responses to climatic stimuli. For symbolic behaviour to constitute linguistic communication between rational agents, it must provide the means for arguing in support of a contention and for denying it; it must, that is, have the equivalents of 'because' and 'not'. If it has any concept of rationality (which the word 'because' implies), it must have some concept of universality. Hence it must allow not only for the present tense 'It is snowing', 'It thunders', but also for omnitemporal clauses 'Whenever it snows ...', 'Whenever it thunders'. And if these admit of denial, it must be possible to cite past incidents in support. Hence, even if we used only impersonal verbs, we should need to refer to dates, which would function in the logician's sense as particulars.

Not only would dates be things we could, and would have to, refer to, but with the aid of the date-quantifier 'whenever' (or, equivalently, 'never') we would be able to introduce other things to refer to. If we can say 'Whenever there is lightning, there is thunder' (or 'It never happens that both there is lightning and it does not thunder'), we have the generic concept of thunderstorm, and with a rudimentary dating system and under reasonably benign climatic conditions we can individuate thunderstorms enough to refer to a particular one. The quantifiers, as Russell saw, give us an effective method of referring: we would be able to refer not only to events, but to any other concomitances that interested us, and in particular to secondary (e.g. chemical) substances (e.g. $(x) [x$ is yellow $\& x$ is ductile $\& x$ is fusible $... \supset ...])$ and material objects (e.g. the golden circle that was visible yesterday). We can, sufficiently for the purposes of communication, if not in a completely watertight way, use descriptions to characterize what it is we want to talk about. And therefore, if we are able to talk at all, we must be able to talk about things, at least in the wide sense of 'what can be referred to by means of a referring expression'.

If we have the wide concept of a thing, we shall be under pressure to develop a narrower concept too. Once we are liberated from the interminable monologue about present climatic conditions, there will be a choice of what to talk about, a choice not completely under the control of speaker or hearer, and hence things talked about will be felt by each party to discussion as being independent of him, and potentially resistant to his will. On the other hand, as we have seen, the need for intersubjective indentifiability will make whatever

[2]   See Jonathan Bennett, *Rationality* (London, 1964), §§ 6–9, pp. 49–86.

is invariant over time and space, as well as over persons, pre-eminently suited to be a thing. And finally, the need for re-identifiability generates a *nisus* towards a certain degree of permanence, although not an absolute one.

§ 21
The argument from change to things

Things, in the wide sense, do not necessarily involve change. We can be good Platonists and confine our discourse to forms and numbers and other mathematical entities. We can. But mathematical discourse is both limited and in a special sense otiose. As Plato himself showed in the *Meno*, in mathematics we do not essentially depend on the testimony of others. The most that other people can do for a mathematician is midwifery. Anything that is true in mathematics is in principle accessible to any mathematician on his own unaided by any others. It is only the weakness of the intellect and not the nature of the subject that makes a mathematician want to listen to other men. Ideally, the mathematician sees himself as a solitary soul, perhaps communicating his results to the learned world, but not needing to receive anything from other mathematicians except acclaim. Mathematical communications being in this sense otiose cannot contain, in a corresponding sense, anything new. Two hundred and fifty-seven can never cease to be a prime number. It makes no sense to suggest, or to deny, that it might. The very unchangingness of mathematical entities is what is responsible for there being no news value in mathematical communications, and thus in one sense no need for them.

Communications that are not, in this extended sense, otiose, must convey propositions which are in some sense contingent. The negation of the proposition must have been not so obviously impossible that there was no point in ruling it out. And if, moreover, we are to regard the communication really as news, there is implicit the possibility of things being different now from what they may have been once. There is thus a weak argument from things to change. Not everything we can talk about need admit of change, but only those things that do – or at least might – change will be talked about much.

The argument in the reverse direction, from change to things, is much stronger. If there is change, there is something that changes. For if there is change we can describe it, at least to a minimal extent; and if we can describe it, we must be able to say both what the difference is, and what it is that, in spite of the difference, makes us think that we are referring to the same thing. Once we allow that there is more than one topic of discussion, it is not enough to deny what we previously had affirmed or to affirm what we previously had denied. For there to have been change, there must be not merely a change of thing talked about, but something, the same thing, that has changed. The verb 'to change' cannot be an impersonal verb, like 'to thunder'. If there are changes, then there are things that change.

§ 22
The argument from things and change to different sorts of qualities

If we have things changing we must distinguish between some qualities that characterize a thing essentially and others that are applicable only *per accidens*. For in order to identify the thing before it changed and re-identify it again afterwards we must be able to describe it to some extent, so that both the speaker and the hearer can talk about the same thing. Therefore it must have some qualities that it has both before and after the change. But equally there must be some quality that it has before but not after or after but not before the change, in virtue of which it is said to have changed – that is, to have become different. The former qualities are said to be essential, the latter only "accidental". The distinction between essential and accidental qualities need not be the same for all things, nor indeed need it be the same for one thing on all occasions. Nevertheless it would be nice if we could have one pervasive and permanent distinction.

§ 23
Qualitative identity and numerical distinctness

Tweedledum and Tweedledee were alike in all respects. There was nothing that could be truthfully said of Tweedledum that could not be truthfully said of Tweedledee. Tweedledum was fat: so was Tweedledee. Tweedledee wore a schoolboy cap: so did Tweedledum. Tweedledum was quarrelsome, but timid in the face of unorthodox ornithological monstrosities: exactly the same could be said of Tweedledee. Yet, although one in corpulence, in dress sense, and in temperament, they were two people. They were the same in respect of any quality we cared to consider, but different when we came to count them. We express this technically by saying that Tweedledum and Tweedledee were qualitatively identical (the same in all respects) but numerically distinct (counting as different when being counted).

In some cases the concept *qualitatively identical but numerically distinct* is clearly inapplicable. With numbers, it makes no sense to think of two different numbers being alike in all respects. If they are alike in all respects – if they have all the same factors, say – then they are the same number. There is only one number that is divisible by 2, 3, 4 and 6 and by no others (apart from 1 and itself), namely 12, and there is only one number that is next after 12, namely 13.

We cannot think of numerically distinct numbers being qualitatively identical. With numbers, the rule is: if distinct, then different.

So too with theories, so too with tunes. If we have two tunes that are note for note, pitch for pitch, tempo for tempo, the same, then they are not two tunes but just one tune, the same tune in each case. So too with angels and chemicals. Angels, according to the Schoolmen, could not copy Tweedledum and Tweedledee. Every angel was a type: every specimen was a species. If there are two angels, there must be two species of angels, since each one is a species all on its own. Each one is the unique – the logically necessarily unique – member of its own species. Angels are even more exclusive than electrons – there cannot be two angels in exactly the same "state" – that is, having all their characteristics exactly the same. In each state, or species, there is logical room for only one, which then, logically, excludes all others. Exclusive occupancy is enjoyed, too, by chemical elements of their places in the modern periodic table; and, more generally, chemical substances – secondary substances, as Aristotle would call them – cannot be qualitatively identical without being the same substance. I cannot have carbon$_{12}$dum and carbon$_{12}$dee with all their properties the same but nevertheless constituting different substances. If they are to be different substances, then there must be some difference in their properties – different melting point, different refractive index, different atomic number – to make them different. Else they are the same. The elements in the periodic table are paradigm examples of secondary substances: they cannot be qualitatively identical but numerically distinct, and, by the same token, they cannot change. It makes no sense to talk of carbon$_{12}$ changing its properties. Plato would be pleased at the static structure of theoretical chemistry. If the fundamental furniture of the world consists of Platonic forms or secondary substances of which it makes no sense to talk of two being qualitatively identical but numerically distinct, then it must be impossible also to talk of things changing. For if a thing can change, it may do so with respect to a quality so remote, trivial and accidental that it remains the same in all essential respects: and hence, if we consider the two cases, in one of which the thing has changed and in the other of which the thing has not changed, we shall be considering two things that are qualitatively identical but numerically distinct. Thus the possibility of change implies the applicability of the concept qualitatively identical but numerically distinct, and *per contra* the latter is a necessary condition of the former. Plato's belief in the forms is thus connected with his sense of the unreality of time. For Aristotle temporal processes are very real; but then for him secondary substances were definitely secondary. If our ontology gives pride of place to secondary substances, then we shall view the world in some such way as did Plato or Leibniz or Bradley, and shall be inclined to say that time is unreal [1], or that it is only an image of eternity, or

[1] F. H. Bradley, *Appearance and Reality* (Oxford, 1930), ch. IV, pp. 33–6, ch. XVIII, p. 181.

that, although a well-founded idea, it is not anything more, and does not enter into the fundamental account of things which we ought to give if we are speaking as truthfully as we possibly can [2].

[2]   G. W. Leibniz, Fifth letter to Clarke, § 47; reprinted in H. G. Alexander (ed.), *The Leibniz-Clarke Correspondence* (Manchester, 1956), pp. 69–72; see also *New Essays*, II, 13, 17, and elsewhere.

§ 24
Types and tokens

We have shown that if there are things that can change, it must be possible to apply the concept of some things being qualitatively identical but numerically distinct. We can argue for this possibility from a different starting point, that of the means of communication.

The type–token distinction is basic to all methods of communication. We can grasp it best by considering answers I might give to the question 'How many words?' If I were asked 'How many words there were written in the previous section, and I made a count, I should come to the problem: Am I to count the word 'quality' on p. 112, l. 8, and the word 'quality' on p. 113, l. 29, as one word or two words? Sometimes we feel quite clear that they should be counted as two words; sometimes, equally clearly, we feel that they should be counted as one word. When writing a book, it is quite clear that everytime I use the word 'quality' I am using another word. If, however, we are wondering 'How many words there are in the English language?' then it would be a poor joke to add to the score every time some one opens his mouth. We resolve this difficulty by talking in the first case about *word-tokens*, in the second case about *word-types*. If each occurrence of a word counts separately, we are concerned with word-tokens; if all the occurrences of the same word count only for one, we are concerned with word-types. When sending a telegram it is word-tokens I am charged for; when confessing my ignorance of the German language, it is word-types I admit to knowing so few of.

We can draw the same distinction between token and type with books, musical notes, and whole musical works. Thus 'How many books are there in my room?' means 'How many token-books?': when I am asked how many books I have written in the last five years, I should not give some answer like '6000'; I am being asked 'How many book-types?' Similarly in music, if I play a note – middle C – and I play the same note again, I have played two middle C note-tokens, but only one middle C note-type. Or again, every time I play a long-playing record of Beethoven's Fifth Symphony, I have as it were a new performance-token of this particular performance-type. Indeed, there are a number of different performances of Beethoven's Fifth Symphony, some by the Hallé and some by the Royal Philharmonic Orchestra, and if I was extremely keen on Beethoven's Fifth Symphony, I might have several long-playing records of it, all of the same symphony, but not, in another sense, all of the same performance of the symphony. And thus we have a double sense of the type–token distinction: if I possess the Toscanini Beethoven Fifth Symphony and the Barbirolli Beethoven Fifth Symphony, each of these is a type, *vis-à-vis* the number of *playings* of each long-playing record; e.g. if I wonder whether I should change my sapphire needle, I am concerned with the number of playings of the record. But though we contrast the record with playings of the record, and say that playings of the record were in some sense performances and that records represented, rather, a type of performance, yet we should regard these same records as tokens *vis-à-vis* the more generic type, Beethoven's Fifth Symphony. We have, as it were, two performances of the same symphony. And we draw a contrast between the friend who has thirty-five long-playing records which are different performances of the same symphony, with another more normal friend whose thirty-five records are all performances of different musical works.

Although the type–token distinction can in this way be applied elsewhere, its paradigm application is in language. For a language must consist of words or symbols, so understood that different speakers on different occasions can utter or recognize some sound or shape as being the *same* word or the same symbol. Without the possibility of using the same word on different occasions, it would be impossible ever to classify different things as possessing some common quality, or to refer to the same thing on different occasions; that is, language would be impossible. If we are to have any means of communicating with one another, we must be able to regard *different* sounds or shapes on *different* occasions as being the *same* word or symbol: that is, we must regard them as being different word-tokens of the same word-type. In learning to read we learn to count

not   not   Not   NOT   *Not*   NOT as all being the same word-type. We learn to ignore the idiosyncracies of my handwriting. We learn that it does not matter whether words are handwritten or printed, in lower case letters or capitals or italic, written in red ink or green or blue, in illuminated gothic or in serif type. Only some geometrical (indeed, usually only some topological) properties are important for the written word. Similarly, in learning the spoken language, we learn to discount such "accidental" qualities of the sounds as whether they are uttered by a man or a woman, quickly or slowly, crossly or calmly, in the Queen's English or in some local dialect; but we pay attention to the accentuation and the length of vowels, and learn to count as different words 'not' with a short 'o' and 'note' with a long 'o'. The length of the 'o' is an essential quality of a word, whereas the timbre, etc., are only accidental qualities. So far as language goes, different word-tokens of the same word-type are qualitatively identical although numerically distinct, and it is only because we can find application for the concept of sounds or shapes being qualitatively identical but numerically distinct that we can communicate at all. Thus the argument from the possibility of communication to the concept of qualitative identity but numerical distinctness.

§ 25
The Identity of Indiscernibles

The concept of two things being qualitatively identical but numerically distinct is not only a difficult one, but a doubtful one. Leibniz, and many other philosophers, maintained the doctrine of the Identity of Indiscernibles [1], that if allegedly two things were alike in absolutely all respects without there being any differentiating feature between them at all, then they were not two things but just one and the same thing; if they are really qualitatively identical they are numerically identical as well. After all Tweedledum and Tweedledee did have different names, are represented as occupying different portions of space, and, if identical twins, were born one slightly before the other. These differences are not essential: they are differences of personal *fiat*, of spatial and of temporal position – just the differences that the pressure of communication towards Platonism leads us to discount (see § 13, pp. 75–6). They are typically *in*essential qualities. But they are enough to make a difference, to make *the* difference between them. It is not an essential quality of Tweedledum that he should be sitting on the right of the bench, and of Tweedledee that he should be sitting on the left. But it is enough to enable us to discern the one from the other.

[1]   G. W. Leibniz, Third letter to Clarke, §§ 2–6, Fourth letter to Clarke, §§ 21–6; reprinted in H. G. Alexander (ed.), *The Leibniz-Clarke Correspondence* (Manchester, 1956), pp. 25–7, 36–7, 61–3; see also pp. xxiii–xxiv.

Leibniz's doctrine of the Identity of Indiscernibles is very difficult to discuss. Every formulation seems to be inadequate, and both parties seem to have missed the point of the other's argument. This is partly due to the confusion between what we can conceive and what we can know (see § 5, pp. 27–8; § 11, pp. 67–8). We can conceive of Tweedledum and Tweedledee changing places, although we (that is, all of us apart from Tweedledum and Tweedledee themselves) could never tell whether this had happened. Our concept of symmetry depends on this contrast. If I look at, say, a regular pentagon, I can see that if it were rotated through 72° it would look exactly the same as it does now; but for that very reason I cannot tell whether it has been rotated through 72° or not. It is only because there is a certain operation I can think of – rotation through 72° – that makes no difference, and leaves the whole figure absolutely the same, that I ascribe this particular sort of symmetry. And so with all other symmetries. A similar resolution applies in another difficult case, that of two fundamental particles – say two electrons – where we cannot possibly tell which electron is which, but have good theoretical reasons for saying that there are two electrons, not one. We can conceive that there are two electrons, not because we can count them, but because our theory admits the possibility, and distinguishes the case of there being two electrons from that of there being only one. But although we can, in a non-vacuous fashion, conceive of there being two electrons, we cannot know which is which. Our knowledge does not – and cannot – enable us to tell anything about one of them rather than the other. They are epistemologically indiscernible, but not conceptually identical.

Nevertheless, Leibniz had a point. Although we can talk of being qualitatively identical but numerically distinct, if we are to apply this description in practice, as we often do, we must do so in virtue of differentiating between essential and inessential qualities, and understanding qualitative identity to refer only to the former. It is in this weak sense that the possibility of things changing and the possibility of communication both imply the applicability of the concept qualitatively identical but numerically distinct. But when we come to do metaphysics we want to strengthen this sense as far as we can. We are speaking of the fundamental things in the universe, and want to say of them whether they can be qualitatively identical but numerically distinct. It does not matter for monists; they think there is only one fundamental entity, as did Spinoza. But it is an acute problem for pluralists, who think that there are many, numerically distinct, things, and must either make them out to be all qualitatively different or else find some schema for avoiding the full force of the Identity of Indiscernibles.

It depends on what counts as a quality. In metaphysics we are likely to go to extremes. We may be generous, and allow every quality to count as a possible characteristic of, and indeed as an essential quality of, the fundamental things; or we may be sparing, and not allow any quality to characterize a fundamental thing, not even as an inessential quality. In the one case, every substance has an infinite number of qualities (positive or negative), and however many qualities two substances have in common, there will be some further qualities which one will have and the other not. Two substances may be highly similar – they may have many, important, qualities in common. But the question of their having all qualities in common cannot arise, because we could never count all their qualities to see that they did each have them all. In the strict sense, there-fore, we could never say of any two things that they were qualitatively identical, although speaking more loosely we might call two things qualitatively identical if they had all important qualities in common; and then we should say that they were qualitatively identical though numerically distinct because, although each had the same important qualities as the other, there were other, relatively unimportant, qualities possessed by the one but not by the other. What these unimportant qualities were we could not say in advance. Presented with a pair of "qualitatively identical" substances we could find some respect in which they differed, but it could be a different respect in different cases. There would be no uniform "quasi-quality" which always served to distinguish qualitatively identical substances that were numerically distinct.

Leibniz's monads are such substances. They, like persons, are infinitely complex. We never can exhaustively describe them, and therefore, however similar they are, they are never the same in all respects. There will always be some further feature to differentiate between two monads or two persons, however much the same they seem to be at first. People are different. They resemble one another in many ways, but in different ways. No two people resemble each other in all respects, if we can attach meaning to those last two words, and we do not feel any need for some uniform quasi-quality to differenti-ate between people otherwise totally alike. Each man is unique, and though some may seem at first sight to be all alike, it is only a question of knowing them better to discover their peculiar characteristics whereby each is himself and not anybody else.

Newton and Locke take the opposite course to Leibniz, and give as bare a description as possible of their fundamental substances. Their atoms or corpuscles or particles are thought of as things – material objects – rather than as persons, and things have no souls, no personality, no originality, to make each one itself alone. They are not unique. Even with everyday material objects, we often have no hesitation in regarding them as qualitatively identical; and much more so with the metaphysical substances of Newton and Locke, which must be qualitatively identical since they none of them have any qualities whereby they could be different from any other. Locke's "something I know not what" is, confessedly, as uncharacterized as anything could be. It is, in the material mode of speech, the analogue of the existential quantifier in Russell's theory of descriptions. Substances may on occasion be described, but substance in itself is devoid of qualities and is completely characterless. If to this minimal account of substances we add, as Locke and Newton did, the belief that there are many substances – many fundamental things – then the problem of differentiating between them becomes acute. A monist can afford to say little, but a pluralist must at least say enough about his substances to show why he thinks they are more than one. The Newtonian atoms cannot be shown to be different from one another in virtue of any qualities some do and some do not possess, for they all possess none. And therefore, if they are to be distinct, there must be some quasi-quality, which is not enough of a quality to prevent them from being all qualitatively identical, but is enough to enable them to be numerically distinct. And this is the reason for space.

§ 26
Parameter space

The argument of the last section can be put in a less metaphysical form. The principle of the Identity of Indiscernibles makes us attend to the distinction between essential and inessential or "accidental" qualities. We may regard this distinction as being always merely relative or as being really absolute. That is, qualitative identity may always be merely relative; or there may be some absolute standard of qualitative identity, according to which things really are qualitatively identical, not by recourse to counting certain qualitative differences as irrelevant to the classificatory purpose in hand, but by having clearly in mind what are, and what are not, qualities irrespective of any classificatory purpose we may have.

If the first alternative holds, then the difference between qualitative difference and numerical difference is only one of convenience. Any two things that are numerically different will also be qualitatively different in some way or other, although the difference may be too insignificant for us to want to count it *as* a qualitative difference. But it could be counted as a qualitative difference, and therefore, viewed absolutely, it should be counted as a qualitative difference. There will be, strictly speaking, no *infima species*: every specimen will be the sole member of its own species. If we have only monadic qualities – if, that is there are no relations (dyadic qualities, because relations obtain between *two* things) that can be used to differentiate between one thing and another – then there will have to be an infinite number of combinations of these monadic qualities if it is to be possible for there to be an infinite number of individuals. It may be that this is the right conceptual scheme to apply to individuals that are rational, sentient agents; because although rational sentient agents (e.g. persons) can have relations with one another, we sometimes feel that each person is unique and qualitatively different from every other, quite apart from any relations he may have with any other.

A very different example of each specimen's being the sole member of its own species can be given provided that one dyadic quality is allowed, namely a one–one asymmetrical relation whose domain is the whole universe of discourse. This then will be exemplified by the natural numbers or the integers. Each natural number is qualitatively different from every other one. Although we can of course have classes of natural numbers, each possessing some common quality (e.g. the even numbers, the perfect squares or the prime numbers), nevertheless each number is unique and differs from every other number. It would be absurd to talk of two numerically distinct numbers being qualitatively identical.

Let us now consider the second alternative, that two numerically distinct individuals really can be qualitatively identical, and not merely by disregarding some qualitative difference on the particular occasion. We shall need some way in which things can differ without being qualitatively different, some absolute, uniform quasi-quality as we have called it. Furthermore, this way of being different needs to be continuous, because change is continuous, and one of our arguments for introducing a non-qualitative way of being different was in order to allow for the possibility of a thing's changing – that is, the same thing being different at different times. We thus have, in effect, a *parameter*, assigned to each individual, which distinguishes one individual from others qualitatively identical with it, which can take different values at different times, and which can vary continuously. This is what constitutes the philosopher's space. It is room for things to be different in; different from one another at any time, different from themselves at different times; and because in a sufficiently short time things can differ by as little as we please, space must be, like time, a continuum.

§ 27
Wireless metaphysics

The medieval disputations about the number of angels that could dance on the point of a needle were not as silly as modern philistines make out. For angels, being incorporeal entities, could not be individuated by the positions of their bodies in space. Without the quasi-quality of spatial position to distinguish one angel from another, there was no way for angels to be qualitatively identical but numerically distinct. Hence, each angel had to be the sole specimen of a whole angelic species. Thus the very incorporeality that prevented one dancing angel pushing another out of his way on the top of the needle prevented also there being more than one angel, of any particular species, to occupy that vantage point. Angels, along with all other spiritual principalities, are forbidden by Pauline injunctions from multiplying their kind.

Our need for space, to give us room to be different in, has nevertheless been denied by many philosophers. Strongly though he argues in favour of the primacy of spatio-temporal bodies, Professor Strawson believes it is possible to have a no-space world [1]. I shall attempt to show how, even in his purely auditory world, there will be stronger analogues to things and to space as I have introduced it than any he will allow; and in doing this I shall carry the

[1] *Individuals* (London, 1959), ch. 2, pp. 63–86.

conclusion of the previous section one stage further, and show that the para-meter space required for things to be different in must be a space of more than one dimension.

My purpose is different from Strawson's. He is concerned primarily with the problem of solipsism; space is important to him as providing room for things to continue to exist in even when not present to consciousness. I am concerned primarily with the problem of individuation; the importance of space to my argument is that it allows similar things to be different in spite of all their similarities. Also, I shall propose a slightly different model from Strawson. He gives, as an analogue to his model, the tuning of a wireless set. Let me make this my actual model. That is, I shall allow to my occupants of a purely auditory world one kinaesthetic sense, that of rotating a control knob (rotation, it should be noted, contains no dimension of length: it is, according to the usual reckoning of physicists, a pure number; in any case, our supposi-tion is of kinaesthetic sense of a simple scalar quantity) in order that they may be agents as well as passive observers [2]. Each listener can "move" in "wave-length space" – or, if it be objected that the concept of *wave length* presupposes that of *length*, let it be called, equivalently, "frequency space" (frequencies have dimension T^{-1}). And, since he can know whether he is making a movement (though not necessarily where he is), without auditory sense experience, I have no need of Strawson's 'master-sound' [3]. Strawsonian purists and hi-fi enthusiasts, however, can consider, instead of the wireless model offered, one constructed on the basis of several variable speed tape-recorders, where the key note of each piece replayed performs the role of Strawson's master-sound.

A third difference between Strawson's approach and the one adopted here lies in the use of the type–token distinction. Strawson appears to regard it as an absolute one [4], whereas I have taken it to be a relative one (see § 24, p. 116): a particular long-playing record is a type *vis-à-vis* the playings of it on a record player, a token *vis-à-vis* that issue of records as listed in catalogues and discussed by record critics. If the distinction is absolute, and a criterion of distinctness of token is temporal discontinuity, then there are bound to be insuperable difficulties in *re*-identifying sound-tokens. But in fact these sound-tokens are not the only possible particulars in an auditory world, any more than glimpses or sights of material objects are the only possible particu-lars in a visual world. In the wireless world we are imagining we can regard either stations or programmes as particulars. We can distinguish between Droitwich and Wrotham, and can re-identify either of them after having listened to the others; and equally, we can switch from Radio 3 to Radio 2

[2]   Compare *Individuals*, ch. 2, p. 83: "The idea of oneself as an agent forms a great part of the idea of onself."
[3]   ibid. pp. 76–7.
[4]   ibid. pp. 69–71.

and back to Radio 3, and identify again the talk on Chinese epigraphy that had led us in the first place to twirl the knob.

Not only are programmes or stations possible candidates for auditory thing-hood, but they, or something like them, are the only possible ones. For things must continue to exist in time: they must be, at least relatively, permanent. Else they could not be the subject of change; and, although we can talk of timeless entities, such as numbers, and impermanent ones, such as events, communication is in a certain sense otiose with respect to the former, and would be difficult to establish if the latter were the only sort of thing. Thinginess implies permanence – witness the belief of the materialists that atoms are everlasting [5]. We need not pitch our demands as high as that – only a relative permanence is required. But that is enough to rule out purely transitory sounds as paradigm things. Things in a purely auditory world must continue long enough for us to be able to recognize them as being the same things as we heard on some previous occasion, in spite of having in some respects changed. Voice would do, or any characteristic pattern of sounds: in New York, the random listener can always recognize WQXR – the New York Times Station – because it is always playing Beethoven symphonies.

Although we have argued that things must continue to exist, they do not need to exist continuously, but only continually. It is necessary that there should be some later dates at which we can recognize the thing as the same thing, although having changed; but it is not necessary, so far as the present argument goes (but see § 28, p. 130), for the thing to have existed at all intervening dates. In a purely auditory world a succession of discontinuous sounds would be acceptable – it is quite acceptable that stations should be off the air for a period – provided there was some recognizable pattern that recurred. Often, in fact, there is repetition – the tune introducing "Music While You Work" or "The Archers" on the B.B.C., or repeating the First Symphony immediately after the conclusion of the Ninth on WQXR. But if Beethoven were alive and had emigrated to America, he might well be retained by the New York Times to write a new opus every day, in which discerning critics could recognize the hand of the master without any obvious repetition.

Wireless programmes differ from wireless stations in two important characteristics:

(i) They can be in two "places" at once.

(ii) They cannot be qualitatively identical but numerically distinct.

We can get the same programme on different wavelengths (or frequencies) – for example, the B.B.C. Radio 3 on 464 m and 194 m. And if we get a programme at different 'places' it does not count as two different programmes unless there is some qualitative difference between them. They are to this extent more universal than wireless stations, more like secondary substances. If a meta-

[5]   See, for example, Lucretius, *De Rerum Natura*, I, 150, 215–16, quoted above, § 13, p. 74.

physician of the wireless world were to take programmes as his fundamental things, he would be taking the Leibnizian alternative. Programmes are like monads in having a sufficient degree of complexity to enable them to be different in an indefinite number of ways. Every programme would be highly characterized, each unique and unlike every other in important respects; and wavelength space, although well founded in kinaesthetic experience, would play no vital role in his account of fundamental things.

If, on the other hand, we take the stations option in our wireless metaphysics, we shall get not monads but atoms. Stations are innately characterless; they can, and occasionally do, switch programmes, and they could conceivably all broadcast the same programme. This is how the wireless successor to Newton and Locke conceives them; his monolithic conceptual structure is an Orwellian nightmare in which every wireless station is always reading the same extracts from Pravda in monotonous Russian. Some stations actually are, and all Orwellianly might be, qualitatively identical although numerically distinct. And the two linked ways in which programmes and stations differ reflect the role that is being played in the station analogy by wavelength space, the role of enabling two stations to be different in spite of not differing in any programmatic detail. Even angels, although not differing in the messages they bring, and singing all together in unison, could still be many, provided only they all kept to their separate places in the heavenly host.

The role of enabling things, although qualitatively identical, still to be different is an equivocal one. On the one hand, a difference in spatial position is *eo ipso* enough to make a difference: on the other, a difference in spatial position does not constitute a qualitative difference, and therefore all spatial positions are qualitatively the same. In view of the former, we are inclined to regard space as something absolute, since a difference of spatial position *per se* is enough to make a difference; but in view of the latter, we are impelled to the conclusion that space is in some sense relative, since a difference in space by itself is not to count as constituting any difference at all. When we think about space, we think of it rather as we suppose God does. Every point is immediately present to our mind, and can be named by a number (or ordered set of numbers). But when we try to apply our conception we have difficulty in telling what number (or ordered set of numbers) should be assigned to any given point, because there is nothing inherent in space, according to our present conception of it, that would require us to assign this number (or ordered set of numbers) rather than that, Each thing is in its place, but all places are alike. The only way we can discern one place from another is by appeal to things. And then our concept of space is no longer absolute but relative to those things we take as points of reference [6].

Many difficulties arise from the barely compatible requirements we expect our concept of space to satisfy, and the contrast between what we can conceive

[6]   Compare the similar difficulty over time; see above § 5, p. 28, § 11, pp. 67–8.

of and what we can actually know. Sometimes we are impelled to think of space rather positively, as Aristotle did, and Newton in his metaphysical moments, and as in the General Theory of Relativity today. There space is a quasi-*quality*, and other qualities can be explained in the last resort by reference to space, which appears to be something absolute, with intrinsic properties all of its own. At other times we want to strip away all qualities from space, giving it a minimal role in our conceptual structure and no role at all in our schemata of scientific explanation. In so far as we do this, physics ceases to be a form of geometry, all explanations are in terms of something other than space, and we are at pains to say how much a *quasi*-quality it is.

§ 28
Impenetrability

The wireless metaphysician who takes the stations option is able to be an uncomplicated pluralist, with an adequate supply of substances combined with a strict economy of qualities. But he has not, as yet, accommodated change. Stations are naturally stationary; all the changes we normally experience in our auditory experience are ones that the minimum quality metaphysician is inclined to discount. But in fact stations do move. Often we have to keep turning and counterturning the knob to keep ourselves tuned in to the station we want to hear. And we could conceive this happening much more freely than it does now. But not indefinitely much more – at least not if Orwellian conditions obtained. We could not have under Orwellian conditions what we sometimes now experience, when one station "passes through" another station, and we have stations – identified by their characteristic programmes – in the opposite order to their normal one in wavelength space. For if all stations had the same programme there would be no way of distinguishing them except by their position in wavelength space. And if they can move in wavelength space, not even position is enough to re-identify a station. It might seem that having divested stations of all distinguishing qualities, and having now made even the sole surviving quasi-quality inadequate for re-identifying, we had destroyed this essential condition of thinginess. It is here that we invoke continuity, in the strong sense not required in the previous section (p. 126). Continuity involves only time and space, and therefore is available even in an entirely characterless world in which the fundamental things have no qualities, but only quasi-qualities or parameters. We have already seen that since time has the order-type of the continuum, there must be a parameter with the order-type of the continuum too, to allow for continuous change. We now reverse the argument and make continuous change of parameters a criterion of identity. Although the same thing can be in different places at different times, it will be the same thing only if the positions it occupies at all intervening times together constitute a continuous path from the one place to the other.

From this it follows that although one station can occupy different positions in wavelength space at different times, and different stations can occupy the same position in wavelength space at different times, different stations cannot occupy the same position in wavelength space at the same time. For if they came to occupy the same position at the same time there would be no way of telling them apart, or identifying them thereafter. In our actual everyday wireless experience we can distinguish the following two cases when stations 'wander' in wavelength space:

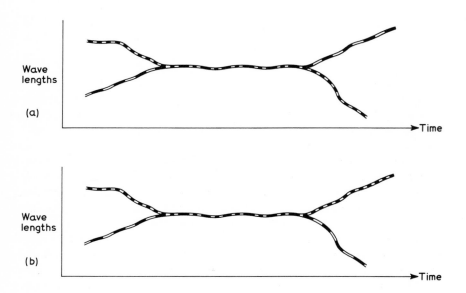

But if there are no differences in programmes – if all stations are a mono-chromatic Pravda grey – then we cannot distinguish the two cases (a) and (b), because all we shall have to go on is the undifferentiated case:

We shall have no ground for considering (*c*) to be a case of (*a*) rather than (*b*), or vice versa. More, we shall not be able even to ask the question; (*a*) and (*b*) are not only indistinguishable, but indescribable.

It follows that (*c*) must not happen. If we are to talk about fundamental things, which are self-subsistent characterless entities, differentiated only inasmuch as they are differently located in space, we must insist that different things always are differently located in space. Else, they will cease to be different. It is a requirement imposed by us, not on us. We can think of many entities that diverge or converge or both. The life history of an amoeba could be represented thus:

and we can enter into difficult disputations whether this should be described throughout as one thing, occupying latterly two places at once, as above; or two things, occupying formerly both the same place, thus:

or as one thing and its offspring, thus:

or as one thing becoming two other things, thus:

We get comparable difficulties describing the life history of paramecium, a protozoon like amoeba, but rather more vegetable-like, in which fission is preceded by fusion. A more everyday example is given by the Ministry of Transport's classification of roads. We are quite happy to see

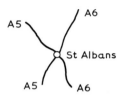

or

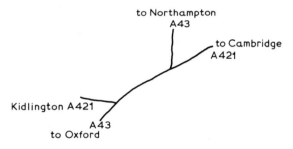

But then we do not regard roads as substances or an amoeba as a fundamental thing. Much ingenuity has been devoted to formulating a logic that does justice both to our intuitions of thinginess and to these everyday examples, but so far without success; and recently an eminent logician, who has made the most determined attempts to do so, has concluded that "the only way in which the ordinary logic of identity can be fully preserved is by maintaining that cases of this sort never occur or can occur, i.e. that it never is or can be the case that one individual thing becomes two individual things" [1]; nor, we should add, vice versa; nor both. Hence impenetrability.

We now see why Locke listed impenetrability as a primary quality [2]. Many have thought that it sits ill with position and velocity, but have been inclined

[1]   A. N. Prior, "Time, Existence and Identity", *Proceedings of the Aristotelian Society* (1965–6), p. 189; reprinted in A. N. Prior, *Collected Papers on Time and Tense* (Oxford, 1968), pp. 78–87.
[2]   *Essay Concerning the Human Understanding*, bk II, ch. 8, § 9.

to reckon it among the secondary qualities or powers, along with ordinary hardness or solidity. But it is not ordinary solidity. It is a conceptual, not a physical, hardness. It is really none other than the Identity of Indiscernibles, in the material mode of speech.

Wireless metaphysics remains defective. Although things can move, they cannot move very much, on account of impenetrability. With impenetrability the order of stations remains invariant, however much stations together shift along in wavelength space. Each station is sandwiched in between two neighbours, and cannot get past them and be next to any other stations. And this is inevitable with impenetrable things located in a one-dimensional space. But it runs counter to our general, non-wireless, experience. Things not only can change, but can change in their relations to one another: they are not arranged in one invariant serial order. It follows that space, philosopher's space, has more than one dimension.

We have still to show that more than one dimension need be continuous. We could, it has been suggested [3], meet the impenetrability argument by having a second discrete, two-valued dimension. Space would then be represented by two as it were parallel lines, a front street and a back street so to speak: if I cannot get past a neighbour in the street, then when she is in the front street, I slip through the house and make my escape down the back street. And indeed we could thus economically surmount the impenetrability obstacle; but at the cost of discontinuity. The change in the discrete-valued space would necessarily be discontinuous. And we need continuity as our last remaining criterion of identity. We could abandon it only if we could offer some other criterion. It would be all right if we could survey the whole of space all at once, God-like, and note the sudden disappearance of one thing from one half of space and the simultaneous appearance of a thing in the other half; then it would be reasonable to interpret it as the same thing going from one side to the other. Or we could manage with a less panoramic survey if we could establish local correlations between parts of the two halves of space. But these would be difficult to establish inasmuch as every thing was either in the one half or in the other, but could never have a foot in both camps. Things do not have feet. But minds do. As soon as we conceive a world that is not a world inhabited exclusively by things but one that contains people too, we no longer need rely for our criterion of numerical identity on spatio-temporal continuity alone because memory and personal characteristics provide other criteria of identity which could suffice if continuity broke down. I could be in one world and remember having been in the other, for memory, like all consciousness, is even in our ordinary life punctuated by the discontinuity of sleep. Or again, I could recognize in the one world a friend's face, handwriting or literary style, which I had become acquainted with in the other; for these are highly characteristic, and are adequate in practice, and perhaps in principle too, to identify

[3] By Mr A. Slomson of Merton College, during a lecture.

uniquely with no possibility of reduplication; and neither vary continuously, nor need to in order to secure identity. Slomson's space, if suitably enlarged and peopled by persons, sets the scene for Quinton's myth [4], which is intelligible just because everyone can have a foot in both camps in a way in which no thing can. A discontinuous parameter can be introduced into a world of featureless things only at the cost of considerable complexity. But complexity is just what the wireless metaphysician seeks to avoid. He wants to have his fundamental substances as simple as possible. But so long as his metaphysics is based exclusively on things – so long as he takes the stations (or atoms) option, not the programmes (or monads) one – he cannot have a radical discontinuity, or he loses numerical identity altogether. He therefore has to have the second dimension of space a continuum of order-type θ like the first.

[4]   A. M. Quinton, "Spaces and Times", *Philosophy*, XXXVII (1962), pp. 141–44. For further discussion, see R. G. Swinburne, "Times", *Analysis*, XXV (1964–5), pp. 185–91; R. G. Swinburne, "Conditions of Bitemporality", *Analysis*, XXVI (1965–6), pp. 47–50; A. Skillen, "The Myth of Temporal Division", *Analysis*, XXVI (1965–6), pp. 44–7; K. Ward, "The Unity of Space and Time", *Philosophy*, XLII (1967), pp. 68–74; Martin Hollis, "Box and Cox", *Philosophy*, XLII (1967), pp. 75–8; F. R. Stannard, "Symmetry of the Time Axis", *Nature*, CCXI (1966), pp. 693–5; Martin Gardner, "Can Time go Backward?", *Scientific American* (January, 1967), pp. 98–108; Martin Hollis, "Times and Spaces", *Mind*, LXXVII (1967), pp. 524–36; J. J. C. Smart, "The Unity of Space-Time", *Australasian Journal of Philosophy*, XLV (1967), p. 214–17.

§ 29
Dimensions and continuity

The number of dimensions of a continuous space is, like continuity itself, a "topological invariant" of that space. Any reasonable redescription of the space must use the same number of terms as the original one. We can define the *dimension-number* of a space as the number of coordinates needed to specify completely, but no more than completely, the "position" of a point. It is, so to speak, the number of *questions* we need to ask about a thing – a substance – in order to be able to know what state it is in; or, again, the number of different ways in which a thing can *alter*. Or we can define it in terms of *boundaries* – roughly, the boundary of an *n*-dimensional space must be itself only (*n*−1)-dimensional. Mathematicians have used all these approaches [1], which can be shown to be equivalent [2]. We shall concentrate on the "*questionnaire*" aspect, although we have implicitly appealed to Menger's

[1] For a general survey see W. Hurewicz and H. Wallman, *Dimension Theory* (Princeton, 1948), ch. I. See further L. E. J. Brouwer, "Beweis der Invarianz der Dimensionalzahl", *Mathematische Annalen*, LXX (1911), pp. 161–5. See also H. Lebesgue, *Mathematische Annalen*, LXX (1911), pp. 166–8. H. Poincaré, *Revue de Métaphysique et Morale*, XIX (1912), p. 486. L. E. J. Brouwer, "Über den natürlichen Dimensionsbegriffe", *Journal f. Math.* (Crelle), CXLII (1913), pp. 146–52. K. Menger, *Dimensionstheorie* (Leipzig, 1928). P. S. Alexandroff, "Dimensionstheorie", *Mathematische Annalen*, CVI (1932), pp. 161–238.
[2] P. S. Alexandroff, op. cit.; W. Hurewicz and H. Wallman, op. cit., p. 24.

boundary definition in our argument that space must have more than one dimension. We shall consider how many questions any scheme for labelling points of a space must ask in respect of any point if it is to label it adequately and economically. It has been shown that the number of questions is independent of the particular labelling scheme adopted, and depends only on the space. Provided we have some notion of nearness or of open set, and hence of continuity, any redescription of a space will need to use the same number of parameters, coordinates or terms, if it is to preserve continuity and is to be neither redundant nor inadequate.

If we drop the requirement of continuity, we can easily redescribe space with fewer, or more, parameters. For example we can redescribe a two-dimensional space – the positive quadrant of a plane – as a one-dimensional space – the positive segment of a line – by the following method. Let the coordinates of any point in the plane be (a, b) where a and b are real numbers. Let the decimal expansion of a be ... $a_5 a_3 a_1 \cdot a_2 a_4 a_6$ ... where $a_1, a_2, a_3, \ldots$ can take the values 0 to 9, and similarly the expansion of b be ... $b_5 b_3 d_1 \cdot b_2 b_4 b_6$ .... Let c be the point $(c, 0)$ on the X-axis, where c is the real number

$$... b_7 a_7 b_5 a_5 b_3 a_3 b_1 a_1 \cdot a_2 b_2 a_4 b_4 a_6 b_6 a_8 b_8 ....$$

It is evident that there is a one–one correspondence between all the points of the form (a, b) and all the points of the form $(c, 0)$. Also that all the points of

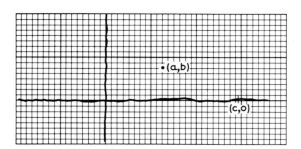

the plane are of the form (a, b), and that all the points of the form $(c, 0)$ lie on just one line. Hence we have a one-dimensional redescription of the points on a plane which is adequate and non-redundant. But continuity is lost. However close two points (a, b) and (a', b') are, the corresponding points $(c, 0)$ and $(c', 0)$ may not be close. For example, on the vertical line $x = 1$, on the lower side of the point $(1, 1)$, we shall have, as the correlate to $(1, 1)$ the point $(11 \cdot 0, 0)$ whereas the correlates to all points on the line below $(1, 1)$ will be of the form $(1 \cdot 0909 \ldots, 0)$. The same decimal expansion that makes the one–one correspondence possible, by coding two numbers as one number, also injects a separation in the coded numbers whenever there is a change of digit in either of the two coordinates of points on the plane. But this will always be so. For whenever two numbers are distinct they differ in respect to some digit. Therefore a continuous line in a two-dimensional space has corresponding to it a discontinuous set of points in the one-dimensional space.

We can also keep continuity provided we accept some degree of inadequacy or redundancy. Obviously we can redescribe part of a two-dimensional space in a one-dimensional way, preserving continuity, for we can do that much with any line in the plane. Much less obviously, we can correlate a two-dimensional space on to a one-dimensional one, without loss of continuity, by means of a one–many correspondence. Peano constructed a continuous curve that went through every point in a given square – but went through many points more than once [3].

It is only when we insist that our redescriptions are neither inadequate nor redundant, but redescribe each point in just one way, and that continuous curves shall be redescribed as still being continuous, that dimension-number appears to be invariant. But this is the sort of redescription that is important for us. We are obviously not interested in inadequate descriptions, nor, less obviously, in ones that give more than one description to any point – for then we lose our criterion of difference; and continuity is of crucial importance in establishing the identity of things. Hence the importance of redescriptions of this sort – topological transformations, as mathematicians call them – and hence also of dimension-number.

[3]   G. Peano, *Mathematische Annalen*, XXXVI (1890), p. 157.

III The theology of space

§ 30
Newtonian space

Newtonian space was the correlate of Newtonian things. "In the beginning", Newton rewrote Genesis 1, 1, "God created atoms and the void" [1]. It is as much a philosopher's space as a physical space, and many criticisms of Newton have therefore missed the mark. Newton distinguishes time, space, place and motion into "Absolute and relative, true and apparent, mathematical and common". Our common notions are, he concedes, relative. "Because the parts of space cannot be seen ... therefore in their stead we use sensible measures of them ... instead of absolute places and motions we use relative ones ..." [2]. "And if the meaning of words is to be determined by their use, then by the names time, space, place and motion, their sensible measures are to be understood .... It is indeed a matter of great difficulty to discover, and effectually to distinguish, the true motions of particular bodies from the apparent; because the parts of that immovable space, in which those observations are performed, do by no means come under the observation of our senses" [3]. Newtonian mechanics, in fact, is entirely relativistic as regards time, space,

[1] *Principia*, Scholium to Definition VIII; reprinted in H. G. Alexander (ed.), *The Leibniz-Clarke Correspondence* (Manchester, 1956), p. 152.
[2] ibid. § 4 (Alexander, p. 152).
[3] ibid. § 4 (Alexander, p. 159).

and uniform motion in a straight line (see below, § 42, p. 197, and § 47, p. 237).

In his philosophical disquisitions, Newton is concerned with the conceptual role, rather than the metrical properties, of space. Absolute space is the concomitant of atoms, corpuscles, absolute things, which enables them to be different and to become different. Space is where atoms *can be*. The universe (in one sense of the word) is where they are. The void (or vacuum) is where they might be but are not. Or, picking up the argument of the previous sections, we may say that parameters are what individuate one atom from another; space is the range of possible values of these parameters; the universe is the set of actual values of all the parameters; the void is the complementary set – i.e. the set of possible, but non-actual values. From this it follows that we cannot talk of space moving, because it is itself the possibility of motion. Speaking in the material mode of speech we might say, with Newton, "that the primary places of things should be movable is absurd" [4], or, more modernly, the range of a variable is itself invariable. But Newton would have been clearer if he could have copied Ryle and said that the range of a variable was itself neither variable nor invariable; and that space was neither movable nor immovable, neither mutable nor immutable.

Since space is a possibility of existence, it is not an existent itself [5]; since it is where things can be, it is not itself a thing. It has some of the characteristics of thinginess – it is intersubjective, independent of time and space, and in some ways recalcitrant to our will – but it has been conceived as playing a complementary role to things, and is therefore something insubstantial and unthinglike. Leibniz was right to deny that space was a substance [6], although the argument he adduces tells only against the confused version of the doctrine of absolute space. Absolute space may be absolute and real, but it is not a fundamental thing in the Newtonian schema, and therefore in one sense of the word 'exist' does not exist – it does not exist in the way that material objects or atoms do – even if in another sense it does exist. If not wrong, it was at least confusing for the Newtonians to maintain that space is a real absolute being. But Leibniz was in his turn wrong in arguing that, since space is not a substance, a being, an attribute or a quality, it must be a relation, and therefore ideal. For relations are no more ideal than qualities and, more important, the list is not exhaustive. Rather than any of these, we should categorize space logically as being a *manifold*, that is to say a many-dimensional continuum. An n-

---

[4]   Alexander, op. cit., p. 155.
[5]   Compare Barrow: *"Tempus igitur non actualem existentiam, at capacitatem tantum seu possibilitatem denotat permanentis existentiae; sicut spatium capacitatem designat magnitudinis intercedentis."* ("Time therefore denotes not an actual existence but a capacity or possibility of permanent existence in as much as space designates a capacity of intervening magnitude.") Isaac Barrow, *Lectiones Geometricae* (London, 1670), *Lectio* I, p. 3; reprinted in *The Mathematical Works of Isaac Barrow, D.D.*, ed. W. Whewell (Cambridge, 1860), II, p. 161.
[6]   Third letter to Clarke, §§ 2–5 (Alexander, op. cit., pp. 25–6).

dimensional continuous space is a set of all those points that can be characterized by a set of n real numbers, or, more abstractly, it *is* a set of ordered
n-tuplets of real numbers. Philosopher's space is, logically speaking, an
instance of a mathematical space – hence the homonymy, hence also the
confusions. Leibniz's list of categories was too limited. Mathematical space is
neither substance nor quality nor relation; it is connected with them. Russell
and Whitehead have shown how we can build mathematical concepts on
slender logical foundations; but for our present purposes, we should regard
mathematical space as a logical category on its own. Newton's absolute space
was such a space, fulfilling a special conceptual role. It was not a metrical
space (except in so far as it needed some notion of "nearness" in order to
define continuity), but only one possessing certain topological properties.
"*In spatio*", said Newton, "quoad ordinum situs *locantur universa*" [7].

   The critics of Newtonian space, Leibniz (on occasion), Mach and Einstein,
urge an epistemological approach: 'How do we tell whether something is
moving or at rest?' Leibniz appeals to the Identity of Indiscernibles: if the
whole universe moved forward (uniformly) in a straight line it would make no
difference, and since "two states indiscernible from each other are the same
state", it is "a change without any change" [8]. Newton did not feel the force
of the epistemological criticism. He takes a "God's-eye" view of the universe.
God is present "from infinity to infinity" and "governs all things, and knows
all things that are or can be done" [9]. There are no epistemological problems
for God. He is "omnipresent, who ... sees the things themselves intimately
and thoroughly perceives them, and comprehends them wholly, by their
immediate presence to himself." [10] He knows, just knows, where everything
is – more, Newton would say, He knows because He puts it there; God places
each atom in its place by *fiat* of His will – and so knows where it is, because He
knows what He is doing, immediately and without room for any epistemological
problem to arise. As with point-particles, so with point-positions. God, on the
Newtonian view, is a Cartesian geometer, the great graph-maker. He knows
all the points, just as He knows all the atoms, individually, and can call them all
by their names. He has no more difficulty in considering a point-position than
a mathematician has when he says "Consider a point (x', y')". Epistemological
problems fail to worry Newton, because he is thinking of an omniscient,
omnipresent Deity whose characteristic relation with things and with space is
expressed in the imperative mood.

   Leibniz complained that Newton's God was arbitrary: but so is a modern
mathematician. He takes arbitrary points in space and assigns arbitrary systems
of labels; he does not have to be concerned with measurement; space does not

[7] *Principia*, Scholium to Definition VIII.
[8] Leibniz's Fourth letter to Clarke, § 13 (Alexander, p. 38).
[9] *Principia*, General Scholium (Alexander, p. 167).
[10] *Opticks*, end of Query 28 (Alexander, p. 174).

have to be based on quantitative relations; some spaces are metric, but some are not. Yet a modern mathematician cannot be completely arbitrary; for spaces (apart from a few special cases) contain an infinity of points, and mathematicians have only finite naming resources. They cannot name each point separately. They must adopt some system of labelling [11]. This they do normally by adopting a coordinate system. In so far as it is a system it is systematic; in so far as they can choose which system to adopt, it is arbitrary. But they cannot choose any system whatever: various requirements – e.g. of continuity – have to be satisfied, which therefore limit the range of choice. But within that range there is a choice that is essentially arbitrary.

Leibniz objected to arbitrariness not only on epistemological but on logical grounds. Two systems that differed only arbitrarily and inessentially were qualitatively identical and therefore only one system after all. It seems difficult to accept this argument as it stands in the mathematical case, once we have admitted the possibility of an arbitrary choice. For the arbitrary choice makes just the quasi-qualitative difference required to secure numerical distinctness in spite of qualitative identity. The Identity of Indiscernibles fails to apply. Leibniz's appeal to it, however, should be construed not so much as a false application of a principle of logic as a programme for mathematics, that where we have "equivalence classes", instead of considering them separately, we should abstract and consider them together as one abstract whole. Thus Frege and Russell turn our attention from particular twelve-membered classes, such as the class of the apostles or the class of calendar months, to the more abstract *class of all* twelve-membered classes, which, they say, constitutes a definition of the number twelve [12]. The temporal and spatial relations that Leibniz thought were the proper concern of science were ones that were indifferent to any displacements of origin or reversals of axes. And in this he was right. Arbitrary alterations may be possible, but cannot signify. They represent, so to speak, a residual egocentricity in our talk about space, which we cannot avoid, but should play down. If we are to talk about space at all, we must have some way of referring to points, some system of labelling points; we must, that is, have some *frame of reference*, some *coordinate system*. But the choice of *this* frame of reference, like referring to a point simply as *this* point, is tainted with egocentricity (see above, § 4, p. 21). It may help to think of each frame of reference (or coordinate system) as an observer, located at the origin, and using the system as his own grid reference. Of course this is not necessarily so: one observer can use different frames of reference, and different observers can use the same one, and it is not obligatory to think of oneself as at the centre of the universe. But one often does. And it is

[11] Compare the similar problem of labelling instants; see above, § 11.
[12]   Gottlob Frege, *The Foundations of Arithmetic*, tr. J. L. Austin (Oxford, 1950), §§ 64–74; Bertrand Russell, *Introduction to Mathematical Philosophy* (Edinburgh, 1919), ch. 2. See next section, § 31.

a helpful device for bringing out the essential egocentricity implicit in any particular coordinate system to associate it with an observer located at the origin, who uses the system as *his* way of referring to it. We shall, in particular, have recourse to this device when we come on the Special Theory of Relativity (see § 43). For the present we need only draw the distinction between what is arbitrary and egocentric on the one hand and what, arbitrariness notwithstanding, remains unaltered, and therefore must be non-egocentric and in some sense real, on the other.

The interplay between arbitrariness and non-arbitrariness constitutes the theme of our treatment of space. We envisage arbitrary alterations, in particular arbitrary redescriptions, of various different sorts, and consider what aspects must nevertheless be unaltered, or what limits on arbitrariness we must impose if we are to preserve some feature we deem to be essential. Newtonian space was not absolutely arbitrary. God could not have created atoms and a discontinuous or one-dimensional void; certain topological properties were required – continuity, connectedness, and some plurality of dimensions. And equally, we cannot describe space except in ways that do justice to these topological properties. We may redescribe space in many, many ways, but not in any way we like. Space, philosopher's space, is thus far defined only "up to" these rather minimal requirements. But they already constitute some restriction on how space is to be thought of or talked about.

§ 31
Equivalence relations and groups

Space can be "described" in many different ways. Geometrical features are those general features of space that are the same however space is described – they are invariant under admissible "redescription" (*automorphism*, to the mathematicians). But how do we characterize a general feature? We do it, as so often in mathematics, by means of an equivalence class.

Equivalence classes are defined by equivalence relations. An equivalence relation R is one that is (in a universe of more than one individual) transitive, reflexive and symmetrical; or symbolically (leaving the variables unbound):

(i) $xRy \& yRz \supset xRz$

(ii) xRx

(iii) $xRy \supset yRx$

An equivalence relation partitions a set into a number of mutually disjoint and conjointly exhaustive classes, each one of which has all its members standing in relation R to all its other members. Each equivalence class has associated with it a quality (*monadic* predicate), which every member of the class can be said to possess. For example, 'being the same age as' is an equivalence relation; the equivalence class is the age group; and the quality that all the members of an age group have in common is their age. Similarly, weight. The fundamental operation is that of balancing. Balancing is an equivalence relation. It partitions heavy objects into classes of objects-all-balancing-against-one-another or, as we say, objects-all-having-the-same-weight. And any particular weight is what is common to some particular class of objects-all-having-the-same-weight [1].

A group is defined as a set of operations (or any other mathematical entities) that satisfy four conditions, namely:

(i') The "product" of two members of the group is itself a member of the group; or, more formally,
(i') If A and B are members of the group then so is AB where AB is the operation that consists of first operating with A and then with B.

(ii') The group "product" is associative; or, more formally,
(ii') If A, B and C are members of the group, then $A(BC) = (AB)C$.

(iii') There is an "identity" member; or, more formally,
(iii') There is a member of the group I, such that $AI = A$ and $IA = A$ for every member of the group.

(iv') Every member of the group has an inverse; or, more formally,
(iv') For every member, A, of the group, there is one and only one other member A^{-1} such that $AA^{-1} = A^{-1}A = I$.

[It should be noted that we do *not* require

(v') The "product" of two members of the group is commutative, i.e. $AB = BA$.

Some groups, however, do satisfy condition (v'); they are called Abelian groups.]

Groups should be compared with equivalence relations. Operations are really one–one relations, whose domains (and ranges, since they always have inverses) comprise the whole universe of discourse. They provide a 'fine-structure', so to speak, of an equivalence relation. The equivalence relation is the disjunction of all the relations that are members of the group. The group provides a set of one–one relations between all the different members of an equivalence class. Conditions (i') and (i), (iii') and (ii), (iv') and (iii), are parallel. In view of (i'), (iii') and (iv') a group of operations can be viewed as an equivalence relation. There is no parallel to (ii'), since any three operations of the group are all viewed as R, so that the parallel to $A(BC) = (AB)C$ would be

$$xRy\&(yRz\&zRw) \equiv (xRy\&yRz)\&zRw$$

which is trivial.

A geometrical property, I suggest, is to be approached through the concept of an equivalence relation. For example, shape. Just as weight is explained in terms of balancing (being the same weight as), so shape is to be explained in

[1] Compare Russell's (and substantially, Frege's) definition of the number 12 as the class of all 12-membered classes. See above, § 30, p. 144.

terms of similarity – being the same shape as. And as a particular weight is an equivalence class of objects-all-having-the-same-weight, so a particular shape is an equivalence class of figures (i.e. sets of points)-all-having-the-same-shape. For example, in Euclidean geometry, circularity is to be explained as the equivalence class of all circles. Any two sets of points that are the same shape, e.g. two circles, have the relation of similarity to each other. This relation of similarity is clearly not one–one. We therefore resolve it into a group, G, of one–one relations – operations – G_r, G_s, etc., so that between any two sets of points that are the same shape there is one and only one of *these* one–one relations between them. If $(a,b,c,...)$ and $(a',b',c',...)$ are the same shape, that is, if aRa', bRb', cRc', ..., etc., or more compendiously

$$(a,b,c,...)\,R(a',b',c',...),$$

then there is some G_r, such that in the usual operator terminology

$$a' = G_r a, \quad b' = G_r b, \quad c' = G_r c, \ldots, \text{etc.}$$

Each member of the group G is an operation which "transforms" the space onto itself; in mathematical terminology, it constitutes an automorphism, and the whole group G is a group of automorphisms. For example, in plane Euclidean geometry, one particular transformation that correlates different figures of the same shape is magnification, say by a factor of two. If we double the size of everything, all figures will be twice as big, but still the same shape. So too if we treble the size of everything; or if we multiply by 6, or $\frac{1}{2}$ or $\frac{1}{3}$. These different magnifications are individual members of the group of all operators that leave shape unchanged. Any two figures that are such that there is *some* member of this group, which when operating on one of the figures will result in the other, are similar (have the same shape) – that is, they have the equivalence relation of similarity obtaining between them – and are therefore members of the same equivalence class.

Every group of automorphisms can thus be construed as a geometrical property – same shape, same size, same topologically – and every geometrical property can be construed as a group of automorphisms. But a group of automorphisms need not be construed as a geometrical property, a set of transformations transforming all the figures with the same property into one another. Instead of this "active" view of a transformation, we can take a "passive" view and regard an automorphism not as an alteration of, or correlation in, a space but simply as a redescription. Either way the formal properties will be the same. It will be a group, and all the results of group theory will apply, whether we regard the automorphism as a transformation of *one* set of points into *another*, or as a redescription of the *same* set of points. And in the argument that follows we shall take both points of view. We shall take one group of automorphisms, and regard some of its members as transformations, constituting an equivalence relation definition of some ‘geometric property, and others as redescriptions, admissible in virtue of some principle of arbitrariness. The bounds of the arbitrary can thus be made to define the figures of geometry.

If we have a group, G, of automorphisms and some redescription, D_s, then two figures $(a,b,c,\ldots)$ and $(a',b',c',\ldots)$ that are G-equivalent in virtue of some operation G_r [so that $(a',b',c',\ldots) = G_r(a,b,c,\ldots)$] will be redescribed as

$$D_s(a,b,c,\ldots) \quad \text{and} \quad D_s(a',b',c',\ldots).$$

If the geometrical property G is preserved under redescriptions of the group D, then

$$D_s(a,b,c,\ldots) \quad \text{and} \quad D_s(a',b',c',\ldots) \quad \text{are } G\text{-equivalent};$$

that is, there is some member of G, not necessarily G_r, but, say, G_t, such that

$$D_s(a',b',c',\ldots) = G_t\, D_s(a,b,c,\ldots).$$

But $(a',b',c',\ldots) = G_r(a,b,c,\ldots)$

$$\therefore G_r(a,b,c,\ldots) = (a',b',c',\ldots) = D_s^{-1}\, G_t\, D_s(a,b,c,\ldots).$$

This is to hold for any sets of points $(a,b,c,\ldots)$, so we need not specify them, and can say simply

$$G_r = D_s^{-1}\, G_t\, D_s.$$

That is, the group G is such that if any member is premultiplied by D_s^{-1} and post-multiplied by D_s, the result is another member of the group: which we express by writing

$$G = D_s^{-1} G D_s$$

or, more symmetrically

$$D_s G = G D_s.$$

A subgroup G of D which is such that

$$G D_s = D_s G$$

for every D_s of D, is called a *normal subgroup* of D. And *per contra* the set of elements D_s for which the above relation holds is called the *normalizer* of the group G. A normal subgroup is, so to speak, one that is, taking the whole subgroup as a single unit, Abelian with respect to every member of the group (see p. 147). In an Abelian group every subgroup is normal, since for every member G_r of any subgroup contained in an Abelian group, D,

$$G_r D_s = D_s G_r$$

and hence $$G D_s = D_s G.$$

In groups that are not Abelian, of course, we do not have to have the same member G_r of the normal subgroup G involved: that is, we do not have to have

$$G_r D_s = D_s G_r \quad \text{for every } D_s \text{ of } D$$

but only, as we had in the first place,

$$G_r D_s = D_s G_t$$

where G_t may be some other member of **G** than G_r.

We have in the mathematical concept of a normal subgroup a purely formal relation between admissible redescriptions of a space and possible geometrical features of it. We can make the argument go either way: you tell me what are the admissible redescriptions, and you have thereby determined also what are to count as geometrical features. Alternatively, I tell you what are the features that are to count as geometrical, and I have thereby laid down limits on what redescriptions are admissible. We shall adopt the former approach, and by deciding what is arbitrary and what is essential in our description of space, we shall discover what its invariant, geometrical character must be.

§ 32
Digression into geometry

Group theory is the best approach to geometry. But group theory involves "operations" (or "operators") and Plato did not approve of operations [1] because operating clearly involved time, and truth – especially mathematical truth – was timeless. In his view, the correct method of doing geometry was the axiomatic method [2]. We need not follow his arguments here – they are not entirely cogent [3]. His ruling was accepted, and even to the present day most thinkers regard geometry done *more geometrico*, that is, axiomatically, as the paradigm intellectual discipline. Mathematicians therefore have come to regard geometries as axiomatic systems. Not all axiomatic systems are regarded as geometries, true: only those that bear some family resemblance to Euclidean geometry, which was the first to be axiomatized. But members of the family are viewed first and foremost axiomatically, as *axiomatic systems*.

[1] *Republic*, VII, 527a. But see *Meno*, 87a, where he had used one of the forbidden words (παρατείνειν) and *Timaeus*, 54–5, where many of the metaphors are dangerously operational. Aristotle follows Plato's precept: see *Metaphysics*, K, 7, 1064a 30; A, 8, 989b 32.
[2] *Republic*, VI, 510b–511d; VII, 527a–534a.
[3] See further J. R. Lucas, "Not 'therefore' but 'but'", *Philosophical Quarterly*, XVI (October, 1966).

In an axiomatic system we have a certain number of primitive symbols and formation rules which specify the well-formed formulae, and among these well-formed formulae a certain number of *primitive propositions*, or *axioms*, say:

$$A1, A2, A3, \ldots An$$

and furthermore certain *rules of derivation*, or *rules of inference*, which enable us to derive a proposition of a certain form, say $\varDelta$, as a conclusion, given one (or more) proposition(s) of certain form(s), say $\varGamma$ (or $\varGamma$ and B, etc., as the case may be) as premisses. Then, if D is of form $\varDelta$, and C of form $\varGamma$, we shall be able to infer D from C. A *proof-sequence* is a series of inferences, where each inference is legitimated by some rule of inference, and the premiss(es) of each inference either is the conclusion of an earlier inference in the proof-sequence or is listed at the outset as one of the premisses required. The conclusion of the last inference is then said to be *derivable from* the premisses given at the outset; or alternatively, we say that the premisses yield the conclusion. We symbolize this by writing

$$P, Q \vdash R$$

where P and Q are the premisses given at the outset and R is the conclusion of the last inference.

We call *theorems* those propositions that are derivable from the axioms alone. If

$$A1, A2, A3, \ldots An \vdash P$$

then P will be a theorem of the axiomatic system whose axioms are

$$A1, A2, A3, \ldots, An,$$

and whose rules of inference are the ones given. It does not matter if only some and not all of the axioms are needed for the derivation. If we can derive P given A1 and A2 as premisses, then so also can we derive P given A1 and A2 and A3 as well. In particular, we allow the trivial derivation

$$A1, A2, A3, \ldots An \vdash A1$$

to count as an inference; so that the axioms count as being theorems in a strict sense.

Although presented as axiomatic systems, geometries are usually not fully formalized. The peculiarly geometrical part is given by the axioms, but a certain background of logic is taken more or less for granted. Euclid lists the following "common notions" (κοιναὶ ἔννοιαι):

Things that are equal to the same thing are equal one to another.
If equals are added to equals, the wholes are equal.
If equals are subtracted from equals, the remainders are equal.

Things that coincide with one another are equal one to another.

The whole is greater than the part [4].

But although he lists them, he does not develop axiomatically the *logic* he uses. We may say, with Church [5], that Euclid's method is an *informal* axiomatic method, inasmuch as the underlying logic is left unanalysed and unformalized.

The boundary between what ought to be axiomatized, as belonging to the special mathematical discipline under discussion, and what can be left un-axiomatized, as being part of the general logic common to all disciplines, is not laid down exactly [6]. Euclid, in fact, took too much for granted, and his proofs rely not only on the common notions of an underlying logic, but on many geometrical and topological intuitions – e.g. of between-ness [7] – which may well be intuitively acceptable but ought certainly to be made explicit. The formal axiomatic method, in which the underlying logic is axiomatized too, avoids this sort of danger, but at the cost of being extremely cumbersome and almost unintelligible. The formal axiomatic method provides a watertight check on hidden assumptions, but the informal axiomatic method gives us more insight into what is really at stake, because there are fewer purely mechanical steps – less dead wood. Euclid was right not to push axiomatization through to the bitter end.

The axioms, together with the underlying logic, determine the axiomatic system; but, in an important sense, the same axiomatic system may be given by various different sets of axioms. There are, for example, many different sets of axioms from which we can derive the propositional calculus, and similarly there are different sets of axioms for Euclidean geometry, or for projective geometry. The criterion of identity for axiom systems is not that they should have the same axioms, but that they should have the same theorems (taking theorems in the strict sense so as to include axioms). Thus if each axiom of one set of axioms is either an axiom of another set of axioms or can be proved from that other set of axioms, then every theorem of the first axiomatic system is a theorem of the second; and if, vice versa, each axiom of the second set of axioms is either an axiom of the first set of axioms or can be proved from that first set of axioms, then every theorem of the second axiomatic system is a theorem of the first. So that if both these hold, we can say that the two different sets of axioms are each *axioms for* or *axiomatizations of* the same *axiomatic system*. In practice, most axiomatizations of any one axiomatic system overlap, and there are only a few axioms of either set that are not axioms of both, and it is only necessary to prove each of these from the other set of axioms.

[4]   Omitting, with Thomas, δ', ϵ', ς' and θ' as possibly spurious and certainly unnecessary.
[5]   Alonzo Church, *Introduction to Mathematical Logic* (Princeton, 1956), vol. I, p. 57.
[6]   ibid. vol. I, p. 58, n. 129.
[7]   For a fuller account see Carl G. Hempel, "Geometry and Empirical Science", *American Mathematical Monthly*, LII (1945), pp. 7 ff.; reprinted in Feigl and Sellars, *Readings in Philosophical Analysis* (New York, 1949), esp. § 2, pp. 239–40.

Euclid's axioms ($\alpha\dot{\iota}\tau\dot{\eta}\mu\alpha\tau\alpha$) were:

1   To draw a straight line from any point to any point.

2   To produce a finite straight line continuously in a straight line.

3   To describe a circle with any centre and any diameter.

4   All right angles are equal one to another.

5   If a straight line falling on two straight lines makes the interior angles on the same side less than two right angles, the two straight lines, if produced indefinitely, meet on that side on which are the angles less than two right angles.

These axioms are, as we have mentioned, inadequate. For example, the assumption Euclid needs, when constructing the perpendicular bisectors of a given line, that two circles, each having an end point of the line as centre, and both having the line as a radius, will intersect, needs itself to be justified. Pasch, Peano, Veblen and Hilbert have developed axiomatizations of Euclidean geometry which include many axioms of order, congruence, continuity and completeness in order to be absolutely watertight. The resulting set of axioms tends to be very cumbersome – further witness to the high price in intelligibility we have to pay in pursuit of absolute rigour. The following set of eighteen axioms is less cumbersome and more intelligible than most [8]. It takes as its primitive concepts *point* and, instead of the more intuitive *line*, *order*. Line is defined in terms of these after the first four axioms have been given; similarly *segment*, which is equivalent to what Euclid would have called a line. The axioms as given here are axioms for three-dimensional Euclidean geometry. We could omit axioms 8 and 9 for plane geometry, provided we then made Desargues's theorem – that if two triangles are centrally perspective they are also axially perspective – an axiom in their stead.

I AXIOMS OF ORDER AND INCIDENCE

1   There are at least two distinct points.

2   If A, B are any two distinct points, there is a point C such that A, B, C are in the order ABC.

3   If points A, B, C are in the order ABC, they are distinct.

4   If points A, B, C are in the order ABC, they are not in the order BCA ....
(*Definitions*: If A, B are any two distinct points, the *line AB* consists of A and B and all points X in one or other of the possible orders ABX, AXB, XAB. The points X in the order AXB constitute the *segment AB*, and are said to lie *between* A and B.)

[8]   Reprinted from Gilbert de B. Robinson, *The Foundations of Geometry* (4th ed. 1959), pp. 51 ff., by permission of University of Toronto Press (© Canada 1940, 1946, 1952, 1959). The more obvious definitions, e.g. of triangle, have been omitted. The reader can omit these axioms without losing the thread of the argument, which is resumed on p. 156.

5 If two distinct points C, D lie on the line AB, then A lies on the line CD ....

6 If A, B are two distinct points, there is at least one point C not on the line AB.

7 If A, B, C are three non-collinear points and D, E are two points in the order BCD and CEA, then there is a point F in the order AFB such that D, E, F are collinear ....

8 If A, B, C are three non-collinear points, there is at least one point D not on the plane ABC.

9 Two planes that have a point in common have a line in common ....

II AXIOMS OF CONGRUENCE

10 If A, B are any two distinct points on a line l, and A' is a point on a line l', then there are two and only two points B', B'', on l', where B', B'' are on opposite sides of A', such that the segments AB and $A'B'$ are congruent to one another, and AB and $A'B''$ are congruent to one another; in symbols:

$$AB \equiv A'B' \quad \text{and} \quad AB \equiv A'B''.$$

Every segment is congruent to itself.

11 Two segments congruent to the same segment are congruent to one another.

12 If the points A, B, C are in the order ABC, and if A', B', C' are in the order $A'B'C'$, and moreover, if $AB \equiv A'B'$ and $BC \equiv B'C'$, then $AC \equiv A'C'$.

13 If BAC is an angle whose sides do not lie in the same line and A', B' are two distinct points, then there are two and only two distinct rays $A'C'$, $A'C''$ from A', where C', C'' are on opposite sides of $A'B'$, such that the two angles BAC and $B'A'C'$ are congruent to one another, and BAC and $B'A'C'$ are congruent to one another. Every angle is congruent to itself.

14 Two angles congruent to the same angle are congruent to one another.

15 If two sides and the included angle of one triangle are congruent respectively to two sides and the included angle of another triangle, then the remaining angles of the first triangle are congruent to the corresponding angles of the second triangle.

III AXIOM OF PARALLELISM

16 If A is any point and l any line not passing through A, there is not more than one line through A coplanar with l and not meeting it.

IV AXIOMS OF CONTINUITY

17 (Axiom of Archimedes) Let A_1 be any point between two arbitrarily chosen points A, B. Take the points $A_2, A_3, A_4, \ldots$ such that A_1 lies between

A and A_2, A_2 between A_1 and A_3, etc., and such that the segments AA_1, A_1A_2, A_2A_3, ... are congruent to one another. Then there exists a point A_n, such that B lies between A and A_n.

18   (Axiom of completeness) The points of a line form a system of points such that no new points can be added to the space and assigned to the line without violating one of the other axioms.

These axioms define Euclidean geometry. There are many other sets of axioms that will define Euclidean geometry equally well [9]; and many other sets again defining other geometries: projective geometry, elliptical geometry (the geometry of Riemann), hyperbolic geometry (the geometry of Bolyai and Lobachevsky), Minkowski geometry, affine geometry and absolute geometry. The relations between the different geometries can be set out in different ways. This is because we count different sets of axioms as axiomatizations of the same geometry provided only the set of theorems in each case is the same. Thus by axiomatizing our geometries in different ways we emphasize different similarities or dissimilarities between them. For example, we can make out Euclidean geometry to be a limiting case between hyperbolic and elliptical geometry, if we consider the sum of the angles of a triangle. The sum of the angles of a triangle in Euclidean geometry is two right angles; in elliptical geometry it is always more than two right angles; in hyperbolic geometry it is always less.

A similar relationship can be made out in terms of the axioms of parallels. We may say that in elliptical geometry there are no parallel lines, in Euclidean geometry there is one and only one line through each point parallel to any given line, and in hyperbolic geometry there are more than one through each given point that are parallel to a given line. This is true. But it gives an over-simplified account of the relationship between elliptical, Euclidean and hyperbolic geometry as they are usually axiomatized. Axiom 16 as stated says only that there is not more than one parallel, but does not state that there is at least one. This is because it follows from the other axioms that there is at least one, and the only issue left open is whether there is only one, or whether there are more. Thus if we replace Robinson's axiom 16 (p. 155) by its negation, 16′ – If A is any point and l any line not passing through A, there are at least two distinct lines through A coplanar with l and not meeting it – we have hyperbolic geometry. If we want to have elliptical geometry, we must alter some other axioms as well as 16. We have, in fact, to alter axioms 3 and 4, the axioms of order, because in elliptical geometry, all lines are closed (as are, for example, great circles on the surface of the earth), so that B's being between A and C

[9]   Hilbert's axioms are given in D. Hilbert, *Grundlagen der Geometrie* (Leipzig), tr. *The Foundations of Geometry* (Chicago, 1902), App. 2; quoted in Anita Tuller, *A Modern Introduction to Geometry* (Princeton, 1967), App. 2, pp. 182–5. See, alternatively, G. O. Birkhoff, "A Set of Postulates for Plane Geometry (Based on Scale and Protractor)", *Annals of Mathematics*, XXXIII (1932); represented in Tuller, op. cit., App. 3, pp. 185–6.

does not preclude C's also being between A and B (as the fact that Singapore is between England and Cape Horn – going eastwards – does not preclude Cape Horn from also being between England and Singapore – going westwards). Thus we need to replace axioms 3 and 4, which give serial order with respect to three points, by an axiom defining order with respect to four points.

Euclidean, hyperbolic and elliptical geometries are all metrical geometries: they have the notion of distance and angle built into them – though in hyperbolic and elliptical geometry the notion seems somewhat paradoxical. If we drop Robinson's axioms of congruence, numbers 10–15, which define equality of distances and equality of angles, and consider only the remaining axioms common to all three, we shall have projective geometry, which deals essentially with rather sophisticated relationships that are the same in whatever perspective they are viewed. Many of the axioms, however, are unnecessarily complicated, being designed to be conjoined with those that have been dropped. For the development of projective geometry alone the following axioms are adequate [10]:

A1   If A and B are two distinct points there is at least one line on which A and B both are.

A2   If A and B are two distinct points there is not more than one line on which A and B both lie.

A3   If A, B, C are points not all on the same line, and D and E ($D \neq E$) are points such that B, C, D are on a line and C, A, E are on a line, there is a point F such that A, B, F are on a line and also D, E, F are on a line.

E0   There are at least three points on every line.

E1   There exists at least one line.

E2   Not all points are on the same line.

E3   Not all points are on the same plane.

E3′   If S_3 is a three-space, every point is on S_3 [11].

Absolute and affine geometry are like projective geometry, and unlike elliptical, Euclidean and hyperbolic geometry, in not starting with a concept of congruence, of equality of distance and of angle, and having therefore no concept of a right angle or of similar triangles. We might therefore classify geometries by dividing them into non-metrical geometries – projective geometry, absolute geometry and affine geometry – and metrical geometries – elliptical geometry, Euclidean geometry and hyperbolic geometry (and also Minkowski geometry, though we shall ignore it here; see below, p. 58, and § 43, p. 205). We then could go on to classify the metrical geometries on the

[10]   Taken from O. Veblen and J. W. Young, *Projective Geometry* (Boston, 1910), vol. I, ch. I, §§ 6–10, pp. 16, 18, 24.
[11]   These axioms again are axioms for *three*-dimensional projective geometry. See above, p. 154.

basis of whether they admit of no parallels, unique parallels, or more than one parallel; or in terms of the sum of the angles of a triangle. Alternatively, we may seek to bring out the greater closeness between Euclidean and hyperbolic geometry than between Euclidean and elliptical geometry. It is natural to do this in the form of a "genealogy" of geometries, starting with the simplest, projective geometry, and then considering the other geometries that can be obtained either by adding further axioms and concepts, or by specializing something already given (as when we introduce the concept of parallelism by designating one of the lines in the plane as a special line, the "ideal" line, sometimes called the "line at infinity"). In this case the axioms of order, which need to be added before the axioms of congruence, will distinguish off elliptical geometry with its cyclic order from absolute geometry with serial order; and the other geometries will be special cases of absolute geometry. The genealogy will then appear as follows:

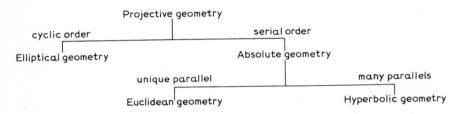

A different closeness of kinship will appear, however, if we drop Euclid's third and fourth axioms – "to describe a circle with any centre and any diameter", and "all right angles are equal to one another" – instead of his fifth axiom of parallels. This yields affine geometry in which the unique parallel plays a leading role, but in which there are no circles, no equality of angles and no similar triangles. Euclidean geometry is then a special case of affine geometry, having these concepts since it is a metrical geometry. Minkowski's geometry is another special case of affine geometry, also having these concepts, but paradoxically a line in Minkowski's geometry can be at right angles to itself – in contradiction to Euclid's own axiom 4. This paradox is allied to the use of complex numbers (that is, involving $\sqrt{-1}$). It is useful for the representation of the Special Theory of Relativity.

§ 33
Interpretations

One of the disadvantages of the axiomatic approach is, as Plato himself dimly discerned [1], that it precludes us from asking the question that most naturally springs to mind – namely, whether the axioms (and hence the theorems) are true. Indeed, from the pure axiomatic standpoint, we cannot even ask what the axioms mean – they constitute an implicit definition of the primitive terms, and give them all the meaning that is relevant to the formal structure of the system. The only things that can be said of an axiomatic system, considered from a purely axiomatic point of view, are to indicate some of the theorems and non-theorems, and whether the axioms are consistent and independent, and, more sophisticatedly, whether they are complete and categorical.

Few mathematicians aspire to be as pure as the absolutely axiomatic approach requires. They interpret. They give meanings to the primitive terms that make the axioms true propositions; from which then it follows that all the theorems are true too. The pure mathematicians visualize; the physicists

[1] *Republic*, VII, 533c: "ᾧ γὰρ ἀρχὴ μὲν ὃ μὴ οἶδε, τελευτὴ δὲ καὶ τὰ μεταξὺ ἐξ οὗ μὴ οἶδεν συμπέπλεκται, τίς μηχανὴ τὴν τοιαύτην ὁμολογίαν ποτὲ ἐπιστήμην γενέσθαι;" ("Where the starting point is something one does not know, and the conclusion and the intervening steps are fabricated out of things one does not know, how on earth can this sort of entailment ever become knowledge?")

seek some physical interpretation. This is how we ordinarily understand geometry. In ordinary life we can more or less recognize a point or a line or a circle when we see one, although only roughly, using various criteria which we normally use but which may not be always compatible; thus our ordinary concept of straight line is given partly in terms of negatives: a straight line is one that has no discontinuities, has no kinks, does not come round back to where it began, has no part of it different from any other part of it (Euclid's "lies evenly with itself"). It is also given in terms of *minimum distance*: a straight line is the shortest distance between two points; which can, granted some true propositions about strings and chains, be exemplified by plumb lines and surveyors' chains. Or again, a straight line can be defined in terms of a light ray: we say that light travels in straight lines and test a "straight edge" by looking down it. Or finally, we can use Newton's first law of motion in the reverse direction, to define a straight line in terms of the path of a freely moving particle not acted on by any force.

Within the limits of our ordinary experience, these criteria agree. We can see that a stone sliding on ice goes in a straight line. The sun, shining through a crack in the shutter into a smoky or dusty room casts a ray of light which we can see to be straight. Alternatively, we could look along a taut cord and see that it was straight in that way; or, again, use a taut cord to discover whether an apparently straight line was really so, or was part of a very slightly curved circle. Beyond the limits of ordinary experience, however, we cannot be sure that the different criteria of straightness will agree; and we must therefore fix upon one criterion as the definitive one, and the others as only contingently, if at all, true of straight lines. Very often physicists take the path of a light ray *in vacuo* as being a straight line; sometimes we use the *geodesic* (minimum distance) property as definite; or we may be able to give precise mathematical sense to "absence of kinks" and define a straight line as a curve whose curvature is zero; or, as in some relativity theories, the path of a freely moving particle is taken to be straight. Each of these provides a different interpretation of a geometry. Under such an interpretation, the axioms, or some of the theorems, will state synthetic propositions which can be put to an empirical test. It then becomes an empirical question whether a particular set of geometrical axioms under a particular interpretation is true or not. If we understand by "straight line" the curve of shortest distance along the surface of the earth between two points on the surface of the earth, then the geometry of the earth's surface – geometry in the original sense of the word – turns out to be not Euclidean but (subject to a small proviso) Riemannian. And similarly, if we interpret "straight lines" as being the paths of light rays, and accept certain well-established parts of physics, then space turns out to be not Euclidean but Riemannian.

Geometry, thus considered, becomes a sort of physics – as its name suggests. Its axioms, taken together, constitute a synthetic proposition, which can in principle be tested, and which makes some significant claim about the world. Granted an interpretation, the question of which geometry is true is simply an empirical one; and, it is often maintained further, only physicists and not philosophers can contribute to its solution. But that is to ignore the role of interpretation in the physicists' concept of geometry. It is only under a given interpretation that a set of axioms becomes determinate, and capable of being tested and found to be true or false. A physicist can always avoid having to reject a set of axioms as false by refusing the interpretation under which they turn out to be false. There may be another interpretation, equally acceptable, under which the axioms turn out to be true. Whether an interpretation is acceptable or not is more a matter for the physicist's judgement than a question to be decided definitely by empirical test. In general, it is possible to secure an interpretation under which a certain set of axioms – say the Euclidean axioms – will come out true, but this interpretation may be purchased at the price of having to have a more complicated physical theory than otherwise. If I interpret straight line as 'path of freely moving particle' and then posit various gravitational and other forces to bring it about that a great many apparently freely moving particles are not really moving freely, I shall be able to maintain that space is really Euclidean, but that physics is a good deal more complicated than Einstein made out. Poincaré was prepared to pay such a price. He thought that non-Euclidean geometries were so inconvenient, and that it was so important to have the geometry on which physics is based Euclidean, that he was prepared to forgo all the simplicity and elegance of the General Theory of Relativity in order to secure its fundamental Euclidean basis [2].

We may have some sympathy with Poincaré. Euclidean geometry does not feel like physics, open to the quick death that is the fate of physical theories once refuted. Euclidean geometry holds a pre-eminent place in our affections, far more so than even the best loved theories of electricity, magnetism or thermodynamics; and although Poincaré is unreasonable in not allowing the General Theory of Relativity to use a non-Euclidean geometry as its basis, we do sense that there is something about Euclidean geometry that makes it less falsifiable than a physical theory, pure and simple, ought to be. Geometrical truths, it seems, are not simply about certain features of our sense experience but rather *schemata*, by means of which we arrange and order our experience. We could call geometrical truths *a priori*, if Kant would soften the necessity of *a priori* propositions to the point of allowing alternative geometries as being at least conceivable [3].

That this is a possible position is secured, as we have seen, by the different

[2]   H. Poincaré, *La Science et l'Hypothèse* (Paris, 1902), ch. III, IV, V, pp. 49–109, esp. pp. 66–7, 79–80, 83; and *La Valeur de la Science* (Paris, 1914), ch. III, pp. 59–95; tr. G. B. Halsted (New York, 1958).
[3]   Perhaps Kant did. See *Gedanken von der wahren Schätzung der lebendigen Kräfte*, § 10 (I 24); and G. Martin, *Kant's Metaphysics and Theory of Science*, tr. P. G. Lucas (Manchester, 1955), ch. I, esp. p. 18.

physical interpretations we may adopt. Within limits we have a choice of interpretations of geometrical concepts, and therefore a choice of the empirical tests a geometry must satisfy. What we have not shown yet is that it is a necessary, or at least a rational, position. We have not shown why we are so anxious to preserve Euclidean geometry, why it seems to us to be intuitively and obviously right and true. Considered simply as an axiomatic system, it is a messy one, far inferior to the abstract elegance of projective geometry, and lacking the substantial metrical and symmetrical merits of elliptical geometry. Nor can we defend Euclid by physics, which neither can support the contention that geometry must be Euclidean nor in fact does that it actually is; and although we may take Poincaré's point, that it is only the conjunction of a geometry and a physics that is testable by empirical observation, so that we always can preserve a Euclidean geometry if we are prepared to pay the price in having a sufficiently complicated physics, yet it would seem altogether irrational to do so Far better, we are told, to abandon Euclid to the historians and the classicists, and to suppress any Euclidean hankerings we may still have, as being merely a carry-over from our schooldays, the vestigial survival of the awe in which we once held our teachers and masters. Nevertheless, the entrenched position that Euclidean geometry holds in our affections is not, as is often made out, mere prejudice or intellectual conservatism, a carry-over from the classical education to which we were subjected in our schooldays. The same sort of arguments which we adduced to establish the topological properties of space (in its fourth sense) can be carried further to show why when we come to consider the geometrical properties of space (in its third sense) we should prefer to have it a Euclidean space if we reasonably can.

§ 34
The measurement of space

Euclidean geometry is a metrical geometry. Length and angle are concepts fundamental to it, and, as we shall see (§ 39, pp. 185–6), in Pythagoras' proposition they are related in the simplest feasible fashion. If, therefore, we can give good reason for having space Euclidean, we have thereby made it metrical. Normally, however, the order of argument is reversed. First a metric is established, and then the claims of the different metrical geometries are considered. There is a danger in this. We do not have a uniform theory or method of measurement. We use very different methods for measuring the temperature of the sun, of a sick child, or of liquid helium; or the distance of the extragalactic nebulae, of a distant church tower, of a length of fabric, of a precision-engineered machine part, or of the radius of a hydrogen atom. Too much emphasis on rigid metre rules is liable to distort our understanding of spatial measurement, not least in leading us to neglect the importance of theory. Measuring, although an operation, is a theory-loaded operation. It is only because we already have a concept of length, related to many other concepts, that we want to measure it, and can devise various methods of doing so.

We have already invoked some notion of nearness as regards space in ascribing to it the topological property of continuity. But as soon as we want to have a quantitative measure of distance, we are faced with the difficulty of comparing distances at different places. How are we to compare a distance here between *A* and *B*, with a distance there between *C* and *D*? We cannot take here to there, any more than we could take a temporal interval today and experience it at the same time as one yesterday. We must have some convention (not necessarily an arbitrary convention) for assigning the same magnitude to different distances in different parts of space. We need to impose a metric mesh that will mark off equal distances in different places and in different directions, just as we needed to pick out a set of instants, with the intervals between regarded as isochronous, in order to measure time.

We often use rulers. We lay off distances using a rigid rod, which, we assume, will be the same length wherever it is, and in whatever direction it is pointing. A clock can be used to measure intervals of time on different occasions, because we think that the causal periodic process – the oscillation of the balance wheel – is independent of date, and that therefore it must have taken as long yesterday for the balance wheel to have oscillated once as it does today. Similarly we use a ruler to measure distances in the belief that the distance between its end points is independent of position and orientation.

We are not obliged to use rulers. Instead of using rulers as the spatial analogues to clocks, we could use clocks themselves, measuring distances by the temporal interval taken to cover them. This is the principle on which radar works, and which is enshrined in the unit "a light year". It depends on there being a speed [1] – as it happens, the speed of light – which is a universal speed – the same always, everywhere, and under all conditions – and which is finite. The first, apparently very strong, condition must, rather surprisingly, be satisfied, granted only rather weak assumptions (see § 42, pp. 198–202). The second, at first sight less stringent, condition is in fact less necessary. It is a fairly contingent matter that the speed of light should be finite; and if the universal speed were infinite, we could not in general carry over our measurements of time to furnish us with measurements of distance too. Although, in the absence of radar we could still use sonar, we should need to have discovered first, by purely conventional methods, what the wind speed was.

Other conventions for assigning magnitudes to distances are possible. In so far as we can adopt an arbitrary system of labelling points, we could adopt an arbitrary system of assigning some distances: two points that differed only in respect of one coordinate would be assigned the distance apart equal to the difference of that one coordinate. But just as we cannot adopt an entirely arbitrary system of labelling points, because there are too many points and we have not time to label them all individually (see § 10), so we are constrained to have a reasonably simple and systematic way of assigning distances – indeed

[1]   I use the word 'speed' to refer to the absolute magnitude of a velocity. That is, to specify a velocity I have to give its direction as well as its speed. My speed is 30 m.p.h.: my velocity is 30 m.p.h. NNE.

we first assign distances, and then use the assignment as a basis for labelling, rather than the other way round. Since space has more than one dimension, we shall need the unit to be given in more than one standard direction. Merely in establishing a metric mesh, there will be a strong temptation to set up a linear space, with a number of unit vectors, spanning the space, as basis, and the rule of scalar multiplication as formalizing an iterated measuring process. When we come further to establish some rule for measuring distances not in any cardinal direction, the pressure to Pythagoreanism will be felt. Even before we reach that point, we are half committed to Euclid's postulate of unique parallels merely by having imposed a metric mesh in which lines never meet. But only half committed. The parallels do not have to be unique; and anyhow, we *can* have a metric mesh – e.g. latitude and longitude – in which some of the lines (those of longitude) do meet – at the North and South Poles. The surface of a sphere is not a linear space. Two different vectors a and b can satisfy the equation $x \times a + y \times b = 0$ without both being null. Thus the argument from having a metric to having Euclidean geometry is much weaker than other arguments we shall now advance.

§ 35
Τὸ ἄπειρον

The argument that space not only must have more than one dimension but must have more than one continuous dimension (in § 28) was not watertight. Its conclusion could have been, alternatively, maintained by an appeal to some principle of parity of esteem between dimensions. Quite apart from the metaphysical argument from the need for continuity as a basis for re-identification, it would be reasonable to demand that all the dimensions of space are *pari passu*, that is, that they all have the same order-type, namely θ. This follows from a deeper requirement, that although the number of dimensions is to be given, the choice of coordinate axes is arbitrary: the space of everyday experience is three-dimensional, but the choice of axes – up–down, north–south, east–west – is arbitrary, and different people at different places on the earth's surface will choose differently, and we could manage perfectly well with, say, northwest–southeast, northeast–southwest, if we wanted to.

We are appealing here to a vague principle that space should be as featureless as possible, just as we earlier appealed to the principle that time should be as featureless as possible. Again, it is as much a *fiat* as a fact; but a purely philosophical and scientific one. The argument about continuous one-parameter groups of transformations (§ 14) does not generalize to groups with more than one parameter. We may be unable to make space homogeneous simply by regraduating its metric. Nevertheless, in our ordinary and scientific thinking we rule that spatial differences, like temporal differences and, in certain circumstances, differences of personal attitudes, are *per se* irrelevant – else no two situations, events or things could be regarded as qualitatively identical and fully comparable [1]. But if spatial differences are to be *per se* irrelevant, then space itself must be undifferentiated, indifferent. Any differences there are between one place and another are to be attributed to something other than the bare fact of their being the places they are – to climatic differences, the incidence of the sun's rays, to historical influences, or the presence of some factor in the one that is absent in the other. But then nothing is left to be attributed to space as such. Space must be featureless, just as substance is, because we have peeled away all possible features, and said that they are not features of space, but of something else. An example will make this clearer. When we throw a die or spin a roulette wheel, we feel that the chance of the die falling with any particular face uppermost or of the wheel stopping in any particular segment must be equal to that of any other one. This we believe not on the basis of a long series of experiments, but because of the spatial symmetry of the system. We see that a cube is symmetrical about all its axes, so conclude that the chance of any one face falling uppermost *must* be equal to that of any other. For if it were not so there would have to be some explanation – e.g. the die was loaded, so that it was not really symmetrical although it looked it. But since, so we believe, it really is symmetrical, then, by contraposition, it really must not be that the chances are unequal. The argument for *a priori* equiprobabilities is a backhanded appeal to the featurelessness of space. And if our *a priori* equiprobabilities are refuted by experience, then we posit an occult factor rather than admit that space has features. If a roulette wheel or any freely moving needle came to rest more often in one direction than any others, then we do not say that that direction is preferred, but that there is some further factor – a magnetic field – to "explain" the apparent asymmetry of space. Thus space is determined to be featureless. And all its properties follow from that. For perfect featurelessness is ineffable. If we can talk about space at all, we must be able to say something about it – indeed, we have shown that it is continuous in more than one dimension, and that its dimension-number, whatever it may be, is an invariant property. And as we shall now see, the very negativeness of its characterization can be made to yield positive properties, much as the negative terms used by the Greek Fathers to describe the Deity were the basis nonetheless of very definite affirmations [2].

[1]   Compare above § 12, p. 72, and § 27, pp. 127–8; see also J. R. Lucas, "Causation", in R. J. Butler (ed.), *Analytical Philosophy*, 1st series (Oxford, 1962), p. 47; and E. Meyerson, *La Déduction relativiste* (Paris, 1925), pp. 106–7.
[2]   G. L. Prestige, *God in Patristic Thought* (London, 1936), pp. 4–10. Note the number of negatives in Henry More's list of divine attributes that are also attributes of space: "*Unum, Simplex, Immobile, Aeternum, Completum, Independens, A se existens, Per se subsistens, Incorruptibile, Necessarium, Immensum, Increatum, Incircumscriptum, Incomprehensibile,*

Time was featureless in two ways. There was first a principle of date-indifference. Any one date was as good as any other date. Therefore the choice of origin was arbitrary. And secondly there was the different principle of indifference as regards intervals in that no natural unit of time, no cosmic year, is given. The metric is arbitrary as regards units, though not altogether as regards the principles of comparison between one interval and another. These two principles carry over to space. We assert a position-indifference, so that there is no natural origin of space; and we assert a unit-indifference, so that there is no fundamental length in the universe. Light years, Angstrom units, and $h \times c^{-1} \times m_e^{-1}$ are equally good, none better, none worse.

With space, however, there are further possibilities of denial. In particular, we assert a principle of direction-indifference. All directions are equal. There are no preferred directions, any more than there are preferred positions or preferred sizes. There is no one proper way to label the points of space. Although the (positive integral) number of (real) numbers needed is fixed, we do not have to have them in the order in which they are given, nor do we have just these particular numbers at all – although, as we have seen (§ 10, p. 63), the assignment of number labels to points cannot be entirely arbitrary. For the sake of simplicity we shall consider the order of coordinates first: that it is only a matter of arbitrary convention that we give the x coordinates first; that it would not make any difference to any spatial properties if we decided to take, say, the y coordinate first; or, more formally, that spatial properties are invariant over interchange of axes. This is really only a special case of a more general principle of direction-indifference: as it were, that there is no reason to take north first; we could equally well take west. But we shall not be content with asserting merely that all cardinal points of the compass are equal – we shall want also to make the stronger claim that we could equally well have coordinate systems, whose axes are not the same as any of those already given, but in between them. To take a simple two-dimensional case, by way of example, instead of having the system OX, OY, we might have the system OX', OY', where OX' is a line pointing *between* OX and OY, and OY' is between OY and the continuation of the X-axis in the direction opposite to X. Thus although the weak thesis does not claim any one direction to be preferred to all others, it does claim a set of "cardinal" directions to be preferred to all other sets of directions. North is no better than south or east or west; but the four points of the compass *are* better than, say, NNE, ESE, SSW, WNW. The strong thesis makes no such claim. Any set of directions, provided there are enough of them, and subject to certain other conditions to be discussed later (§ 36, pp. 174–5), is as good as any other. We should note also that the

---

Omnipresens, Incorporeum, Omnia permeans et complectans, Ens per essentiam, Ens actu, Purus actus." *Enchiridion Metaphysicum* (London, 1671), pt. I, ch. VIII, § 8, p. 69. For Spinoza and Malebranche, too, space was God-like in its negativity: Infinite, Immovable, Indivisible, Unique.

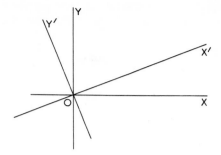

stronger claim allows for a *continuous* shift of axes, whereas any alteration according to the weaker claim must be discontinuous.

Let us, nonetheless, consider the weaker thesis first, since it raises some important issues in a conveniently simple form. In an *n*-dimensional space we need to specify *n* coordinates, x_1, x_2, x_3, ... x_n, in order to specify a point completely. But the space is symmetrical as between these *n* dimensions. It does not matter – so far as the nature of the space is concerned – in what order we specify the *n* coordinates. It is merely a convention. There are therefore no properties of space that would show that one order of coordinates is preferable to any other. So far as properties, general features, of space are concerned, they must be invariant under interchanges of coordinates, although of course the descriptions (i.e. coordinates) we give to particular points will change if we interchange the axes. But general features will not. They will be invariant under this sort of change of axes. What exactly is this sort of change of axes? Any such change is a *permutation*. We permute the order in which we consider the coordinates. If having permuted the order once we permute it again, the result will be another permutation. This may be obvious, but if it is not, let us write the permutation *A*, which re-orders (x_1, x_2, x_3, x_4) [3] as (x_2, x_3, x_4, x_1) in the form

$$A = \begin{pmatrix} 1 & 2 & 3 & 4 \\ 2 & 3 & 4 & 1 \end{pmatrix}.$$

Similarly let us write the permutation *B*, which re-orders (x_1, x_2, x_3, x_4) as (x_3, x_1, x_2, x_4) in the form

$$B = \begin{pmatrix} 1 & 2 & 3 & 4 \\ 3 & 1 & 2 & 4 \end{pmatrix}.$$

Then *AB*, the permutation *A* followed by the permutation *B*, will be

$$\begin{pmatrix} 1 & 2 & 3 & 4 \\ 2 & 3 & 4 & 1 \end{pmatrix} \begin{pmatrix} 1 & 2 & 3 & 4 \\ 3 & 1 & 2 & 4 \end{pmatrix} = \begin{pmatrix} 1 & 2 & 3 & 4 \\ 2 & 3 & 4 & 1 \end{pmatrix} \begin{pmatrix} 2 & 3 & 4 & 1 \\ 1 & 2 & 4 & 3 \end{pmatrix} = \begin{pmatrix} 1 & 2 & 3 & 4 \\ 1 & 2 & 4 & 3 \end{pmatrix}$$

[3]   We consider a four-dimensional space for sake of example.

according to the intuitively obvious rules for manipulation. AB is thus the permutation that interchanges x_3 and x_4 while leaving x_1 and x_2 unchanged. Thus we have it

(i) that the "product" of two permutations is a permutation.

It is easy to verify that the associative rule holds, i.e.

(ii) $A(BC) = (AB)C$.

It is obvious (iii) that there is an identity permutation, and

(iv) that for every permutation A there is another "inverse" permutation A^{-1} which changes the order back again to what it originally was, i.e.:
$AA^{-1} = I$.

It follows, therefore, that the permutations form a group. The group of all permutations of n distinct objects (in our case axes or coordinates) is called the symmetric group P_n: it is particularly important in the theory of groups, as every other finite group can be seen as a subgroup of the symmetric group P_n of permutations of some finite number, n, of objects. We therefore say that general properties in an n-dimensional space must be invariant under permutations of the symmetric group P_n.

The symmetric group P_n of permutations is of order $n!$. For $n \geqslant 2$ it has a subgroup of order 2, and another of order $n!/2$. The subgroup of order $n!/2$ is called the alternating group A_n. It can be defined directly. It is obvious that any permutation could be obtained by a succession of changes in each of which only two axes are interchanged. We call such a change, in which only two axes are interchanged and all the others remain unaltered, a *transposition*. Every permutation then can be expressed as a succession of transpositions. We can bring about a given permutation in different ways, using different transpositions and different numbers of them. Thus

$$A = \begin{pmatrix} 1 & 2 & 3 & 4 \\ 2 & 3 & 4 & 1 \end{pmatrix} = \begin{pmatrix} 1 & 2 & 3 & 4 \\ 2 & 1 & 3 & 4 \end{pmatrix} \begin{pmatrix} 1 & 2 & 3 & 4 \\ 3 & 2 & 1 & 4 \end{pmatrix} \begin{pmatrix} 1 & 2 & 3 & 4 \\ 4 & 2 & 3 & 1 \end{pmatrix}$$

and A
$$= \begin{pmatrix} 1 & 2 & 3 & 4 \\ 3 & 2 & 1 & 4 \end{pmatrix} \begin{pmatrix} 1 & 2 & 3 & 4 \\ 1 & 3 & 2 & 4 \end{pmatrix} \begin{pmatrix} 1 & 2 & 3 & 4 \\ 4 & 2 & 3 & 1 \end{pmatrix}$$

and A
$$= \begin{pmatrix} 1 & 2 & 3 & 4 \\ 3 & 2 & 1 & 4 \end{pmatrix} \begin{pmatrix} 1 & 2 & 3 & 4 \\ 1 & 4 & 3 & 2 \end{pmatrix} \begin{pmatrix} 1 & 2 & 3 & 4 \\ 1 & 3 & 2 & 4 \end{pmatrix} \begin{pmatrix} 1 & 2 & 3 & 4 \\ 1 & 2 & 4 & 3 \end{pmatrix} \begin{pmatrix} 1 & 2 & 3 & 4 \\ 4 & 2 & 3 & 1 \end{pmatrix}.$$

One thing, however, is constant. If we can express a permutation as an *odd* number of transpositions (as here), then we shall always need an odd number of transpositions to express it, whatever the transpositions are and however many of them we use. In this example, A can be expressed by three, or five (or seven, or ...) transpositions, but not by two, four, six, etc. Similarly, if a permutation can be expressed one way by an even number of transpositions, then always we shall need an even number of transpositions to express it.

We thus can divide the permutations of the symmetric group into two sub-sets; those that can be obtained by an even number of permutations and those that can be obtained by an odd number. The identity permutation must count as even, since $I = AA^{-1}$.

For reasons which were originally connected with the theory of determinants, but which will prove to be important also for our purposes, we attach to the even permutations a positive sign, and to the odd ones a negative one. This makes sense if we regard each transposition as having the effect of multiplying by -1. For, if n is even, $(-1)^n = +1$, while if n is odd, $(-1)^n = -1$.

All the even permutations themselves form a group, a subgroup of the symmetric group. For the identity permutation counts as an even one, and the product of two even permutations must itself be even, since it can be expressed as a succession of even cycles followed by another succession of even cycles, and 'even + even = even'. On the other hand the odd permutations do not form a group: there is no unit element, and the product of two odd permutations is not itself an odd permutation but an even one, since "odd + odd = even". The group of even permutations is in fact the alternating group A_n. It is a subgroup of the symmetric group, of index 2 – that is, it is just half as large as the whole symmetric group. But from this it follows that it is a normal subgroup. For consider an element D_s of the whole group, D: either D_s is a member of G or it is not. If it is a member of G, then obviously

$$GD_s = D_s G$$

from the definition of a group (see (i') in § 31, p. 147). If, on the other hand, D_s does not belong to G, then GD_s and $D_s G$ must both belong to the *other* half of D, the set of members of D that are not members of G. For, since G is a subgroup of D, any member of it, say G_r, is a member of D, and therefore the product

$$G_r D_s$$

is also a member of D. But since D_s is not a member of G, $G_r D_s$ is not a member of G. Similarly $D_s G_t$ is not a member of G. But since (as can be easily proved) GD_s has exactly as many members as G, and since there are only as many members of D that are not members of G as there are members of G, it follows that GD_s and $D_s G$ are both exactly equal to the complementary set of G in D, and are therefore both the same. That is, the subgroup G is a normal subgroup of D, and, considered as a single entity, commutes with every member of D.

According to our claim (§ 31, pp. 149–50), normal subgroups formalize our concept of a geometrical feature. They constitute the group of redescriptions under which the geometrical feature will remain the same. If the alternating group is a normal subgroup of the symmetric group, there should be some corresponding geometrical property which it, and not any larger group, preserves. We need therefore to consider more precisely our claim that general spatial features are invariant under any permutation of axes (§ 30, p. 145, and pp. 168–9 of this section). Should we expect general spatial features to be invariant under any permutation of axes, or only even ones? Does the distinction between the alternating and the symmetric group have any geometrical significance? It does. It accounts for the Incongruity of Counterparts, which puzzled Kant, and the concept of parity, which has been of profound significance in modern physics. A left-hand glove is in one sense exactly like a right-hand glove, in another sense totally unlike. They are a pair. They are as like as two peas. But they are different, totally different. It is far easier to get my right hand into your right-hand glove, which does not fit me at all, than into my own left-hand glove, which does. Two gloves are equal: but exact opposites. And this apparent paradox we can explain by distinguishing the symmetric from the alternating group. As regards the symmetric group they are equal – there are permutations of axes that will transform a left-hand glove into a right-hand one and vice versa: but as regards the alternating group they are not in any way similar. The only permutations that will do the trick are odd ones; there cannot be any even permutations according to which a left-hand and a right-hand glove are the same.

Groups provide us with criteria for sameness, and normal subgroups for sameness in respect of some geometrical feature. Since every group has itself as a normal (although improper) subgroup, and nearly every group (every finite group of even order, and many infinite ones) has a normal subgroup half as large as itself, it follows that our concepts of geometrical sameness will show a characteristic bifurcation. There will always be a weaker and a stronger one – similarity and sameness, as we might say – and there is inherent in the concept of featureless space the possibility of two things being similar but evidently, although elusively, not the same. This bifurcation depends not, as Kant thought, on the relation of things to our sensibility [4] but on the mathematical theorem that a subgroup of index 2 is normal, which in turn rests ultimately on the Law of the Excluded Middle.

[4] *Prolegomena*, pp. 286–7, tr. P. G. Lucas (Manchester, 1953), pp. 42–3. See also his *Dissertation on the Form and Principles of the Sensible and Intelligible World* (1770), § 15C.

§ 36
Reflections and rotations

The symmetric and the alternating groups provided two different, though linked, sorts of geometrical sameness. The association of the positive sign with transformations of the alternating group, and of the negative sign with those transformations of the symmetric group that are not also members of the alternating group, accommodates our urge to say 'equal, *but opposite*'. The most typical case of two things equal but opposite is that of reflection in a mirror. And we can see that this does, in our three-dimensional world, correspond to a single interchange – transposition – of axes, if we consider a mirror standing vertically and facing northeast. North then will be reflected into east, and east into north (and south into west and vice versa). Thus it seems reasonable to identify reflections with odd permutations of axes.

There are, however, objections to explaining reflections in terms of inter-change of axes. We naturally think of a reflection not as interchanging two axes, but as reversing the sense of only one axis, leaving all the others un-affected. It was an artificial example to have the mirror facing northeast. We should naturally think of a mirror facing, say, north, and reflecting north into south, but leaving the east–west axis unaffected. More generally, in mathe-matics we should take as the standard reflection the transformation of the point $(x_1, x_2, x_3, \ldots x_n)$ into $(-x_1, x_2, x_3, \ldots x_n)$, not the transformation of it into $(x_2, x_1, x_3, \ldots x_n)$. We picture the mirror as being in the hyperplane of all the other axes, Ox_2, Ox_3, ... Ox_n, rather than in some diagonal one. We then can consider reflection as the simplest case of a more general type of transformation, where the sign of several, and not only one, of the coordinates is reversed. Moreover in one dimension, it does not make sense to talk of interchanging axes, whereas it does make sense to talk of reversing the sign, and we regard doing this as a degenerate case of reflection. We cannot fully meet these objections so long as we consider only the weak claim, which allows axes to be permuted but not to be altered continuously. Only when we consider the more general case does the connexion between interchange and sign-reversal become rationally transparent.

We therefore turn to the stronger claim that there should be parity of esteem not only among the different axes of a given coordinate system, but between different sets of axes. Such a claim, if it is to be reasonable, must be qualified. We must at least insist, as we have already seen (in § 29), that each set of axes should have the same number of axes: else we could not have a one–one continuous transformation from the one set to the other set, and so they could not both describe the same space, identifying every point once and only once, and portraying continuous curves as being continuous. Moreover we must exclude some "degenerate" sets of axes, which, although they seem to give the requisite number of coordinates, are not really, because some of the coordinates can be expressed completely in terms of the others. It would not do, in a two-dimensional plane, to have as one's axes Ox and $O-x$. For, once we had assigned to a point its first coordinate x_1, with respect to the Ox axis, we should have thereby also assigned to it $-x_1$ as the value of its second coordinate. The second coordinate would not really give us any additional information over and above that given by the first coordinate. Similarly in a three-dimensional space the axes are degenerate if they are *coplanar*, all in the same plane. For then, given any two of the coordinates we can work out what the third one must be, and *per contra*, the third one gives us no information about the position of a point in any direction lying not in the plane. In this important sense, therefore, although we seem to have three axes, and have nominally three coordinates for every point, we really have only two. We therefore need to exclude *degenerate* sets of axes or systems of coordinates, and *singular* transformations which lead to such degenerate sets of axes or systems of coordinates. We can express this mathematically by saying that if we have one set of coordinates $(x_1, x_2, x_3, \ldots x_n)$ for a space, a set of n continuous one-valued functions

$$y_1 = f_1(x_1, x_2, x_3, \ldots x_n)$$
$$y_2 = f_2(x_1, x_2, x_3, \ldots x_n)$$
$$y_3 = f_3(x_1, x_2, x_3, \ldots x_n)$$
$$\ldots \ldots \ldots \ldots \ldots \ldots \ldots$$
$$y_n = f_n(x_1, x_2, x_3, \ldots x_n)$$

will give us another set of coordinates and another set of axes, provided the *Jacobian* or *functional determinant* does not equal zero or infinity; that is, provided $J \neq 0$ and $1/J \neq 0$, where

$$J = _{\mathrm{Df}} \begin{vmatrix} \dfrac{\partial y_1}{\partial x_1} & \dfrac{\partial y_1}{\partial x_2} & \cdots & \dfrac{\partial y_1}{\partial x_n} \\[2ex] \dfrac{\partial y_2}{\partial x_1} & \dfrac{\partial y_2}{\partial x_2} & \cdots & \dfrac{\partial y_2}{\partial x_n} \\[1ex] \cdots & \cdots & \cdots & \cdots \\[1ex] \dfrac{\partial y_n}{\partial x_1} & \dfrac{\partial y_n}{\partial x_2} & \cdots & \dfrac{\partial y_n}{\partial x_n} \end{vmatrix}$$

In the General Theory of Relativity we impose few further restrictions on the functions that constitute admissible transformations, but in other cases we make the additional requirement that our space be a *linear space* (or a *vector space*).

A linear space V over a field F [1] is a set of elements, called vectors, such that any two vectors α and β of V determine a unique vector $\alpha + \beta$ as sum, and that any vector α from V and any scalar c from F determine a scalar product $c \times \alpha$ in V, with the properties

V is an Abelian group under addition

$$c \times (\alpha + \beta) = c \times \alpha + c \times \beta \qquad (c + c') \times \alpha = c \times \alpha + c' \times \alpha$$
$$\text{(Distributive Law)}$$

$$(cc') \times \alpha = c \times (c' \times \alpha) \qquad 1 \times \alpha = \alpha \ [2]$$

If we have a linear space, the Jacobian becomes a simple determinant, with constant constituents, and the whole transformation can be represented by a square $n \times n$ matrix, with constant constituents.

Such a matrix is best viewed as a compendious statement of the rules for translating the answers to one questionnaire into answers to another. The coordinates of any point in one coordinate system are the answers for that point to the questionnaire of that system. If we change the questionnaire but still have the answers to the old questionnaire, a matrix will enable us to "process" the old answers so as to provide a new set of answers to the new questions. We put in the old answers, for any given point, along the top, and put in the new questions down the side. In order to get the answer to the first of the new questions, we take the first row of the matrix, and regard the figure in each column of that row as giving us the "weight" to be attached to the old answer at the top of that column; we multiply the entry at the top of each column by the number in the first row of the column, and take the sum of the results, and take that total as the correct answer to the first of the new questions. Similarly

[1]   The field F is in our case the field of the real numbers.
[2]   Taken from G. Birkhoff and S. MacLane, *A Survey of Modern Algebra* (New York, 1953), p. 162.

the answer to the second of the new questions is given by the second row, repeating the same procedure in each column. Similarly the third and all subsequent answers.

	x_1	x_2	x_3	x_4	x_5
x_1'	a_{11}	a_{12}	a_{13}	a_{14}	a_{15}
x_2'	a_{21}	a_{22}	a_{23}	a_{24}	a_{25}
x_3'	a_{31}	a_{32}	a_{33}	a_{34}	a_{35}
x_4'	a_{41}	a_{42}	a_{43}	a_{44}	a_{45}
x_5'	a_{51}	a_{52}	a_{53}	a_{54}	a_{55}

It might seem natural to suppose that if we wanted to reverse the transformation – that is, to recover the answers to the original questionnaire from the answers given to the revised one – all we needed to do was to read off the column entries from the rows – read off $(x_1, x_2, x_3, x_4, x_5)$ from

$$\begin{bmatrix} x_1' \\ x_2' \\ x_3' \\ x_4' \\ x_5' \end{bmatrix}$$

feeding the latter into the rows of the matrix and extracting the former from the columns; or, rewriting this procedure in standard form, we might expect that if the transformation was

	x_1	x_2	x_3	x_4	x_5
x_1'	a_{11}	a_{12}	a_{13}	a_{14}	a_{15}
x_2'	a_{21}	a_{22}	a_{23}	a_{24}	a_{25}
x_3'	a_{31}	a_{32}	a_{33}	a_{34}	a_{35}
x_4'	a_{41}	a_{42}	a_{43}	a_{44}	a_{45}
x_5'	a_{51}	a_{52}	a_{53}	a_{54}	a_{55}

its inverse should be

	x_1'	x_2'	x_3'	x_4'	x_5'
x_1	a_{11}	a_{21}	a_{31}	a_{41}	a_{51}
x_2	a_{12}	a_{22}	a_{32}	a_{42}	a_{52}
x_3	a_{13}	a_{23}	a_{33}	a_{43}	a_{53}
x_4	a_{14}	a_{24}	a_{34}	a_{44}	a_{54}
x_5	a_{15}	a_{25}	a_{35}	a_{45}	a_{55}

The second matrix is called the *transpose* of the second. If a matrix A has elements a_{mn}, its transpose A^* is defined as the matrix a_{nm}: the matrix each of whose rows is identical to the corresponding column of A (and which therefore can be pictured as a "reflection" of A about its leading diagonal). But although the transpose of a matrix can always be defined, it is not in general the matrix of the inverse transformation. That is in general,

$$A^{-1} \neq A^*.$$

We can calculate the inverse, but cannot obtain it by simply reversing the original procedure, treating columns as rows and rows as columns.

In the simple, discrete case, however, in which we only permute the axes, it *is* true that the inverse of a transformation is given by the transpose of its matrix. For in this case, the matrix consists only of 0s and 1s. In each row there will be only one 1, in the column of the coordinate to be given that place in the re-ordering, and similarly in each column there will be only one 1, and the rest 0s; and then, since if $x'_r = x_s$ it follows that $x_s = x'_r$, the inverse transformation will be given by the transposed matrix. Therefore we can say that for permutations of axes

$$A^{-1} = A^*$$

whence

$$AA^* = I.$$

It is this requirement which we generalize from the discrete to the continuous case. A transformation is called *orthogonal* if and only if

$$A^{-1} = A^*,$$

or, equivalently, if and only if

$$AA^* = I \text{ [3]}.$$

Since the determinant of A^* must equal that of A, it follows that

$$|A|^2 = |I| = 1$$

$$\therefore |A| = \pm 1.$$

Those with value $+1$ we shall expect to be the continuous generalization of the redescriptions of the alternating group, those with value -1 we shall expect to include some ineliminable element of reflection. We note that this more general account will accommodate the difficulties mentioned earlier. The matrix $\begin{bmatrix} -1 & 0 \\ 0 & 1 \end{bmatrix}$ for example has determinant of value -1 just as much as the matrix $\begin{bmatrix} 0 & 1 \\ 1 & 0 \end{bmatrix}$, and in particular the 1×1 square matrix $[-1]$ has determinant of value -1, and therefore is a reflection, just as we feel it ought to be.

The transformations that are such that the matrix of their inverse is the same as the transpose of their matrix are the ones we should normally identify as *rotations*, with or without reflection. Those without reflection, whose determinant equals $+1$, represent simple rotations of axes, e.g. N $\rightarrow$ NE, E $\rightarrow$ SE, S $\rightarrow$ SW, W $\rightarrow$ NW; whereas a corresponding one with determinant equal -1 would be N $\rightarrow$ NE, E $\rightarrow$ NW, S $\rightarrow$ SW, W $\rightarrow$ SE (this is only one possibility: it depends on the plane of reflection). We should note that the distinction between involving or not involving a reflection is relative to the number of

[3] See, e.g., G. Birkhoff and S. MacLane, op. cit., p. 258.

dimensions employed. In a plane the two triangles ABC and $A'B'C'$ are reflections of each other, but in three dimensions the one could be transformed

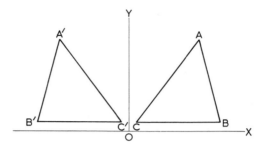

to the other by a simple rotation round the OY axis. The algebraic rationale is that whereas the transformation of a plane into itself represented by a 2×2 matrix is completely determinate, and either has a positive or else has a negative determinant, the transformation of a plane (regarded as a subspace of a three-dimensional space) into itself, represented by a 3×3 matrix, is not completely determinate. If a, b, c, d are given, $\begin{vmatrix} a & b \\ c & d \end{vmatrix}$ is given. But with a 3×3 matrix, we have a choice between

$$\begin{vmatrix} 1 & 0 & 0 \\ 0 & a & b \\ 0 & c & d \end{vmatrix} \quad \text{and} \quad \begin{vmatrix} -1 & 0 & 0 \\ 0 & a & b \\ 0 & c & d \end{vmatrix}$$

If the former is negative, the latter is positive. And hence any reflection can be regarded as a rotation without reflection if an extra dimension is allowed. It is sometimes made a matter of mystery. The Incongruity of Counterparts disappears, we are told, if we take into account the fourth dimension, and a right-hand glove could be made to fit on to a left hand if it could travel in time. But it is a confusion. Time is not a space-like dimension; and even if we did live in a four-dimensional world, although right-hand gloves could be fitted on to left hands, there would be pairs of four-dimensional figures, each the counterpart of the other, but not congruent with it.

We can now see more clearly why time must be, as we assumed earlier (§ 7, pp. 38–9), one-dimensional. If there were more than one dimension of time, and these dimensions were (as is usually assumed) all on a par with one another, it would be possible to have "rotations" in time which would reverse the order in any one dimension of time, just as we can rotate a ruler through $180°$, so that the 12 inch mark is due west of the zero mark instead of due east. A multidimensional space is isotropic as between different directions – e.g. as between east and north – and therefore as between different senses of the same direction – e.g. as between east and west, i.e. between towards-east and away-from-east. But time essentially has a direction, and so is anisotropic. And therefore there can be only one dimension of time [4].

[4]   For a very different argument, see J. Dorling, "The Dimensionality of Time", *American Journal of Physics*, XXXVIII (1970), pp. 539–40, where it is argued that only in a one-dimensional time will protons, neutrons, electrons and photons be stable.

§ 37
The Euclidean group

The orthogonal group represents the admissible redescriptions according to our principle that the orientation of axes is altogether a matter of indifference. If we add to this our earlier principle that the origin is also a matter of indifference, we shall obtain the Euclidean group. The Euclidean group consists of reflections, rotations *and* translations. The properties considered in Euclidean geometry – fundamentally based on angles and distances – are those that are invariant under transformations of the Euclidean group. Euclidean geometry is therefore the appropriate geometry if we are to adopt the principles of origin-indifference and orientation-indifference. If we are prepared to see things from any point of view and from any angle, then the things that we shall see will be things with Euclidean shapes.

Our transcendental justification of Euclidean geometry, however, has two weak links. We have assumed, rather than argued, that space must be a linear space; and we have only a very weak appeal to a sort of symmetry for picking on orthogonal transformations as the natural continuous generalization of the symmetric group. We can put forward some arguments for space being a linear space. Clearly, the requirement that the vectors form an Abelian group – i.e. that for any two vectors, α and β,

$$\alpha + \beta = \beta + \alpha$$

is an extreme case of our requirement of direction-indifference. If all directions are equal then it should not matter whether one goes 1000 miles north and then 1000 miles east, or 1000 miles east and then 1000 miles north; and the fact that on the surface of the earth it does make a difference is one of the indications that in terrestrial geography north is a different sort of direction from east (see § 34, p. 165). The rules for the multiplication of vectors by scalars may then be justified on the grounds that if we were going to have any form of measurement in a space of many dimensions we should have to have these rules. In order to measure time at all, we had to impose a scheme for marking out isochronous intervals; in order to measure space we had better impose a mesh of equal spatial intervals; and then, whatever principle we adopt for assigning measures, the system will constitute some sort of linear space.

These arguments are not watertight, indeed cannot be, or the General Theory of Relativity would be necessarily false. It must be possible to have less stringent requirements of featurelessness – for example, to have only the principle of covariance instead of that of linearity – and still have a workable concept of space. Our arguments are not compelling; nevertheless, they may give rational grounds for *preferring* one sort of space to any other, for *wanting* space to be Euclidean *if* possible, and as Euclidean as possible, if not. If featurelessness is what we are wanting, there is nothing so featureless as what is absolutely flat.

We can produce some further arguments for picking on the orthogonal group. I mention, first and briefly, some purely algebraic ones which are tenuous but capable of further development. Although the fact that the matrix of the inverse of a transformation should be equal to the transpose of its matrix does not, on the face of it, seem to be of peculiar importance, we may seek other properties of orthogonal transformations to justify the importance claimed for them. If we have already defined a concept of distance, we can define orthogonal transformations as those that preserve the absolute magnitude of vectors. If we have already defined orthogonality (or a more general concept of angle), we can define an orthogonal transformation as one that preserves orthogonality. Whitehead gives one purely algebraic characterization of orthogonality or, as he terms it, perpendicularity, in that if we represent displacements by group operators, operators representing two perpendicular displacements are *anticommutative*; that is, $E_2 E_1 = -E_1 E_2$ [1]. Alternatively, we can define both distance and angle (and in particular, orthogonality) in terms of *inner products*. The inner product of two vectors $(x_1, x_2, x_3, \ldots x_n)$, $(y_1, y_2, y_3, \ldots y_n)$ is

$$x_1 y_1 + x_2 y_2 + x_3 y_3 + \ldots + x_n y_n.$$

The length, or absolute magnitude, of a vector is the positive square root of its

[1]   A. N. Whitehead, *New Pathways in Science* (Cambridge, 1935), ch. XII, §§ IV, V, pp. 267–77; reprinted in James R. Newman, *The World of Mathematics* (New York, 1956), vol. III, pp. 1567–73. Note, however, the criticism made above, § 7, p. 38.

inner product with itself. Two vectors are orthogonal if their inner product is zero. Intuitively, we might try to base our concept of orthogonality on that of independence. Two vectors are orthogonal if they are independent of each other. Northeast is not independent of east, because if I go northeast I am going eastwards too; but it is independent of southeast, just because if I move northeast, it makes me move neither forwards nor backwards in a south-easterly direction. If we have more than one dimension of space, we must, by definition of dimension, have some concept of independence; and in a metric space orthogonality is the most natural formalization of it; and the orthogonal transformations are ones under which geometrical properties should be invariant; and therefore geometrical properties should be Euclidean.

Independently of the algebra of vector spaces, the three fundamental operations of reflection, rotation and translation are pre-eminently important, as being the simplest of their respective types. The group whose operator is reflection is the simplest of all non-trivial discrete groups; for if we reflect and then reflect again we are back where we started – that is, the essential structure of the group is given by

$$R^2 = I$$

where R stands for the operation of reflection and I stands for identity. It is evident that this group has the fewest possible basic elements and the simplest possible structure, except for the group whose sole member is the identity operator, which is trivial. The operators of rotation and displacement are both continuous; they differ in that if we rotate far enough to come back to where we started, whereas we can displace further and further without ever coming back. Rotation is a cyclic continuous group, displacement a serial continuous group, and these again are the simplest and most fundamental kinds of continuous group. We can define projective, affine and elliptical geometry (but not hyperbolic geometry at all easily) in terms of other groups, but these other groups are less simple and fundamental than the one based on reflection, rotation and displacement. And therefore Euclidean geometry is, on this score of simplicity, pre-eminent.

§ 38
Shapes and sizes

The Euclidean group has practical as well as formal virtues. It is the natural group for men to adopt who are mobile agents, who see things from different points of view, and move them around from one position to another. We are naturally concerned with those features that are going to be the same from whatever place and whatever angle they are described – that is, those that are invariant under the transformations of the Euclidean group [1].

A very different, more specific, approach depends on the independence of our concepts of shape and size. Elliptical and hyperbolic geometries, although consistent and possessed of many merits, have the demerit of linking the concept of shape with that of size in such a way that we cannot have two figures of *the same shape but different sizes*. It is easiest seen if we consider the surface of a sphere, which is a two-dimensional elliptical space, in which any spherical triangle each of whose angles is a right angle must be an octant of the sphere, and have each of its sides one quarter of a great circle in length. In fact it is a theorem in both elliptical and hyperbolic geometry that if two triangles are similar then they are congruent. Indeed, the negation of this proposition can

[1] See more fully, J. R. Lucas, *"Euclides ab Omni Naevo Vindicatus"*, *British Journal for the Philosophy of Science*, XX (1969), pp. 1–11.

be used as an alternative to the parallel postulate. John Wallis (1616–1703), one of the first mathematicians in modern times to consider Euclid's fifth postulate, showed that it could be replaced by an axiom saying that, given a figure, another figure is possible which is similar to the given one, and of any size whatsoever [2]; and Gerolamo Saccheri (1667–1773) pointed out that it is enough simply to postulate that there exist two unequal triangles with equal angles [3].

We can now carry over the principle of unit-indifference from time to space and argue that only if we have our geometry Euclidean are our units left completely open to our choice. Both elliptical and hyperbolic geometry have a natural unit of area, and elliptical geometry has a natural unit of length as well. A Euclidean geometry has no natural unit either of area or of length, and therefore is less committing, and constitutes a better conceptual framework, just because it forecloses the answers to fewer physical questions [4].

More pragmatically, we may argue that it is a necessary condition of our being able to apply the concept 'same shape though different size' that our geometry should be Euclidean. We might almost say that it was a condition of our having the concept of shape at all – for in a projective geometry, which contains no concept of size, similarity of geometrical configuration would be too general an application to be a reasonable analogue of our concept of shape; while in elliptical and hyperbolic geometry, although there is a concept of size as well as one analogous to our concept of shape, since the two cannot vary independently, it would be unlikely, or at least difficult, for the two to be distinguished in the way we distinguish them. Euclidean geometry, if not an absolutely necessary condition for the existence of the concept of shape, is the only ecological environment in which it, like the concept of rigidity, can flourish and prosper. Thus the price of abandoning Euclidean geometry would be the loss of an important respect in which things can be similar to or dissimilar from one another. We should still of course be able to classify things by colour, by chemical composition, by weight or by specific heat, but we should no longer be able to classify by shape, and this would be awkward; not only would our concepts of area and volume become cumbersome if squares and cubes could not be fitted together to form larger squares or cubes or subdivided into smaller constituent squares or cubes, but it would be conceptually impossible to have scale models, diagrams, maps or blueprints; which would be a pity.

[2]   J. Wallis, "*De Postulato Quinto*", *Opera Mathematica* (Oxford, 1693), II, p. 669–78.
[3]   Gerolamo Saccheri, *Euclides ab Omni Naevo Vindicatus* (Milan, 1783); tr. George Bruce Halstead (Chicago, 1920).
[4]   Contrast the attitude of the General Theory of Relativity, which seeks to provide geometrical answers to physical questions, and therefore prefers a Riemannian to a Euclidean geometry. See below § 47.

§ 39
Pythagoreanism

The argument of the last section is Kantian. We like Euclidean geometry because we are men, and have eyes and hands, and need to operate a concept of shape that will be independent of orientation, distance and size. Rational agents who were not human and did not depend on eyes and hands might not share our preferences. But even disembodied spirits can recognize other, Platonic, merits in Euclidean geometry [1].

The culmination of Euclid's first book of Elements is the proof of Pythagoras' theorem in I, 47. But it would have been better if Euclid had not proved Pythagoras but assumed it, and, taking Pythagoras' proposition as a postulate, proved the parallel postulate as a theorem. It was natural enough when geometry was concerned with laying out the boundaries of the Nile to regard parallel boundaries as particularly important; and graph paper is still important in mathematics, and has provided us with one argument in favour of using Euclidean geometry if we can (§ 34, p. 165; § 37, p. 181). But it is nevertheless a pity that Euclid's genius led him to make the existence of unique parallels his fundamental axiom, for Pythagoras' proposition is a more fundamental one

[1] See Rom Harré, *The Anticipation of Nature* (London, 1965), esp. ch. IV, for a general criticism of such arguments; Harré terms all such arguments "Pythagorean", but it would be confusing if I followed his terminology in this and the next section.

still, and a more distinctive feature of Euclidean geometry regarded as an axiomatic system. It connects the concept of distance with that of a right angle – orthogonality – and it does so in the simplest possible way. If we have any metrical space of more than one dimension we are faced with the problem of how to combine measures in different dimensions; if a place is three miles east of us and four miles north, what distance is to be assigned to the direct route? A straight addition rule (which would give the answer "seven") would be tantamount to a reduction to only one dimension of measurement. A "squares" rule is the next simplest, and fulfils all the conditions we require of any rule for combining measures. In particular, it has the merit (which it shares with formulae of the fourth, sixth and eighth degree) of obliterating distinctions of sign – three miles east and four south will be five miles away just the same as three miles east and four north – which suits the essentially non-negative nature of the concept of distance. Other rules could be adopted, might even be forced on us, but if we have the chance of adopting the Pythagorean rule, no further justification is needed. Mathematicians investigating differential geometries, which are not Euclidean, take care to posit nonetheless that they are "locally Euclidean", that is to say that

$$ds^2 = dx_1^2 + dx_2^2 + \ldots dx_n^2.$$

The concession is, indeed, on a small scale: but could hardly be a larger one.

More fundamentally, we could defend the Pythagorean rule as being the simplest case of Parseval's theorem. Parseval's theorem is concerned with the Fourier expansion of functions (satisfying certain conditions of boundedness or measurability) in terms of cosine and sine functions. The Fourier expansion is

$$f(x) = \tfrac{1}{2}a_0 + \sum_{n=1}^{\infty} a_n \cos nx + \sum_{n=1}^{\infty} b_n \sin nx.$$

a_n and b_n are known as the Fourier coefficients, and are given by the equations

$$a_n = \frac{1}{\pi} \int_0^{2\pi} f(x) \cos nx \, dx$$

$$b_n = \frac{1}{\pi} \int_0^{2\pi} f(x) \sin nx \, dx.$$

a_n and b_n are thus independent of x, though determined of course by f. Parseval's theorem then states

$$\frac{1}{\pi} \int_0^{2\pi} (f(x))^2 \, dx = \tfrac{1}{2}a_0^2 + \sum_{n=1}^{\infty} (a_n^2 + b_n^2)$$

provided that in the interval $[0, 2\pi]$ $f(x)$ is measurable and $(f(x))^2$ is integrable.

We may regard the Fourier expansion as showing how a function $f(x)$ may be plotted in a "phase space" in terms of its Fourier coefficients – its "coordinates" – and a set of fundamental functions – "the axes" of the space. It will be a "space" with a denumerably infinite number of dimensions, corresponding to the basic functions $\cos nx$ and $\sin nx$ ($n = 0, 1, 2 \ldots$), and every function (of a certain type) will be represented by a set of values for $a_0, a_1, b_1, a_2, b_2 \ldots$. And then Parseval's theorem shows that the integral of the square of the function, which we might regard as the square of the vector representing the function in phase space, i.e. the square of the distances from the origin to the point $a_0, a_1, b_1, a_2, b_2, \ldots)$, is, barring slight terminological difficulties with the first term, the sum of the squares of the coordinates. And this is a result entirely uncontaminated by geometrical intuition. The notion of space is an entirely abstract one of independent parameters; the basic trigonometrical function can be defined exponentially; measure theory is purely analytical. So that a mathematician who was so pure as never to descend into geometry, and who had never heard of Euclid or Pythagoras, would still want to have a Pythagorean rule in Hilbert space, and still pay his respects to Euclidean orthodoxy.

§ 40
Theodicy

The arguments for the pre-eminence of Euclidean geometry are neither deductive nor empirical, and therefore will seem odious to mathematicians, who have been brought up to believe that only deductive arguments are valid, and suspect to philosophers, who have been told to think that synthetic *a priori* propositions are logically impossible. To the mathematicians I make no apology. I am not putting forward deductive arguments, and therefore, *a fortiori*, not fallacious ones. Deductive arguments are valuable in their way, but secure their extreme rigour at the cost of being unable to cope with many questions which we should like to see discussed. The modern fashion in mathematics is to put forward some postulates, the need for which is not explained, and derive some conclusions, the interest of which is seldom indicated. Fellow mathematicians, familiar with the field, may be able to supply these deficiencies, but other readers are needlessly handicapped by not being told the point of the exercise, and the writers are deprived of the discipline of having to think out what their "motivation" is, in order to express it clearly in words. Mathematicians usually have their reasons, not deductive reasons but sound ones nonetheless, for adopting the postulates they do adopt and for regarding some results as more worth establishing than others. These reasons need to be articulated. Else Plato's gibe is true (see § 33, p. 159 n. 1), and mathematicians spend their time manipulating meaningless symbols in accordance with arbitrarily chosen rules in order to achieve arbitrarily selected goals. If they are articulated, they can be both communicated and shared, and criticized and assessed. The part played by orthogonal matrices is a central one: we need to say why. Spaces do not have to be Euclidean; but even when they are not, it is usually stipulated that they shall be locally Euclidean. No deductive justification could conceivably be given for such a requirement – a space that is not locally Euclidean is not a contradiction in terms – but informal justifications could be given and should be sought, not only as exercises to clarify the mind but to bring to light hidden assumptions and to suggest new lines of research. The arguments I have given, and shall later give for the three-dimensionality of space (see § 48), are not open to the criticism that they are not deductive – they were never meant to be; but they may well be subject to the more damaging criticism that there are other, better, arguments available. I suspect there are; and would like to know them.

The doctrine that synthetic *a priori* propositions are logically impossible needs to be dissolved as much as denied. At the superficial level at which it is made true by being made analytic, we can offer various ways of taking geometry so that it is not synthetic or is not *a priori* or, indeed, is not a proposition within the meaning of the act. Rather than a proposition, it may be a schema for making propositions, or a prescription for looking for them – in much the same way as the sentence 'Every event has a cause' is sometimes understood. Not every feature of our sense experience is necessarily Euclidean – not apparent shapes for instance; but we discount the non-Euclidean ones, and experience is thus *made* Euclidean by ignoring all non-Euclidean aspects of it. Alternatively, we may fasten on some interpretation, thus ensuring the synthetic character of geometry, and while allowing that some geometries – e.g. that used by navigators – are not Euclidean, maintain that our three-dimensional everyday geometry is Euclidean or very nearly so; and then we may go on to say that it is not a bare contingent truth that it should be so, but one supported by consilience with other general features of experience – if our geometry were markedly non-Euclidean, many other aspects of experience would be altered too.

More fundamentally, we need to reject the sharp dichotomies presupposed by the doctrine that synthetic *a priori* propositions are impossible, and to distinguish various ways in which propositions can be *a priori* and synthetic, and various degrees of necessity and of empirical content that may be involved. We mean by *a priori* many different things. We have argued a series of conceptual necessities, not logical necessities. The reason why we ascribe a certain topology and geometry to space (in its fourth and third senses – see § 16, p. 97) is not that to ascribe any other would be inconsistent or self-contradictory, but that to do so would be awkward, inconvenient or incoherent.

These are not irrefragable grounds. Some sorts of experience may be conceivable which would lead us to ascribe a different geometry or even a different topology to space. But the price would be high; and if we could save the foundations of our conceptual structure by modifying some outlying part, we would, and would be right so to do. *A priori* propositions are ones which we believe in advance of empirical evidence, but are not therefore immune to revision in the light of subsequent discoveries. We anticipate nature, and in our anticipations show our belief that ours is, if not the best of all possible worlds, at least one of the most rational ones. Of course we may be wrong. We may find – as many philosophers and scientists have found before us – that the rationality of the world is revealed in forms we had never thought of; experiment and observation have often led men to principles whose underlying rationality was only later apparent. And the world might be much less rational than it could have been. But if so, many of our fundamental categories of thought would be inapplicable; and, *per contra*, if they are to be applicable, certain conditions have to be satisfied. In our monistic moments we can believe in any sort of unified space-time we like. If we were pluralists, but after the manner of Leibniz rather than Newton and Locke, we should have to have time much as we do now, but could dispense with space, regarding it as an optional extra easily conformed to the requirements of the latest scientific theories. It is only if we believe in persons that we have to have time, and things which they can see and talk about, move and manipulate, and then have to have space with certain topological, and preferably with certain geometrical properties. But these are not absolute nor the only *desiderata*, and we need now to consider certain other approaches to space and time which yield different emphases.

IV Space and time together

§ 41
The plenum

For Newton space was where things could be, sometimes actually were, and often in fact were not. Newtonian space has the conceptual merit of enabling us to identify and re-identify things, especially point-particles, the fundamental things, and of giving us room to move around in, and move things around in, without altering them fundamentally. It has the corresponding demerit of not accounting very well for cause and change. It had to invoke the occult quality of gravitational force. Locke had said that Action at a Distance was a conceptual impossibility, until Newton convinced him it actually existed [1]. And Hume reverted to the view that it was impossible and that "contiguity" was an essential constituent of the causal relation. "I find in the first place," he said, "that whatever objects are consider'd as causes or effects, are *contiguous*; and that nothing can operate in a time or place, which is ever so little remov'd from those of its existence. Tho' distant objects may sometimes seem productive of each other, they are commonly found upon examination to be link'd by a chain of causes, which are contiguous among themselves, and to the

[1] *Essay Concerning the Human Understanding* (London, 1690, 1694, 1695), bk II, ch. 8, §§ 13, 14; compare the fourth edition, dated 1700. See his Reply to Second Letter of Stillingfleet in *The Works of John Locke* (London, 1812), vol. IV, pp. 467–8.

distant objects; and when in any particular instance we cannot discover this connexion, we still presume it to exist. We may therefore consider the relation of *contiguity* as essential to that of causation." [2]

Since, contrary to Hume's view, we think that time and space are continuous, we cannot claim that causes must be strictly contiguous to their effects in either time or space. For in a continuum there are no instants or points that are next to any other ones. We therefore replace Hume's conclusion by his argument "Tho' distant objects may sometimes seem productive of each other, they are commonly found upon examination to be link'd by a chain of causes ...; and when in any particular instance we cannot discover this connexion, we still presume it to exist." Any causal relation between any two events separated by any temporal or spatial interval can be resolved into a set of relations between events separated by smaller intervals which together span the whole interval: that is to say, any causal relation is to be expressed in terms of a differential equation with respect to time and space. Hume's contiguous events give way to ones only infinitesimally apart. The philosophy of causes demands differentials much more than does the philosophy of things, and needs a differential geometry to express it.

Giving up things is difficult. Not only do they play a central role in our ordinary ways of thinking, but they have played a crucial one in our transcendental deduction of space from time. But instead of the argument time–things–space, we can generalize directly from time to a (3+1)-dimensional space-time [3], in which space-time provides an abstract frame of reference, and we characterize the spatio-temporal points referred to by some continuously varying potential or, more sophisticatedly, some directed field. Field theory is the mathematical form of a feature-placing language [4], instead of a particular-characterizing (or feature-thing) one. It is important to note the change in the logical status of space. The space we introduced as the logical concomitant of time and things was a glorified adjective [5]. In answering the question where a particular thing was at a particular date, we were saying something about it. Time and things enabled us to refer: space to describe. But when we amalgamate space with time, the result is, as time was, a glorified analogue of a noun rather than an adjective. We no longer ask where a particular thing is at a particular date, but what sort of field characterizes a particular

[2]   *A Treatise of Human Nature*, ed. L. A. Selby-Bigge (Oxford, 1888), bk I, pt III, sec. II, p. 75.

[3]   Although we make space like time, it is not exactly like. It is a common cause of misunderstanding with the Special and General Theories of Relativity to think of them as working with a four-dimensional space-time, with each dimension exactly *pari passu* with every other one. But the distinction between space-like and time-like dimensions remains, and is essential, even in relativity theory, and is best indicated by our speaking of (3 + 1) dimensions. See further below, § 43, p. 205.

[4]   See P. F. Strawson, *Individuals* (London, 1959), p. 203.

[5]   Compare Clarke's Third Letter to Leibniz, § 3, and Fourth Letter, § 10, in H. G. Alexander (ed.), *The Clarke-Leibniz Correspondence* (Manchester, 1956), pp. 31 and 47; and Clarke's Boyle Lecture, in Samuel Clarke, *Works* (London, 1738), II, 527–30.

point of space-time. We can still accommodate things if we need to, but as adjectives – a special sort of field – rather than as nouns. Dirac delta functions bridge the gap between the plenum on the one hand and the atoms and the void on the other. From the point of view of field theory, the Newtonian world is one where the density of matter is zero almost everywhere, except at those points occupied by atoms, where the density must be said to be infinite in order to accommodate a finite mass in a point of no magnitude. But the legitimacy of delta functions is in doubt; they are the bastard offspring of discrete desires in a continuous matrix, and the field theorist is inclined to dispense with particles. But with the particles go our criteria of individuation, of identification and of re-identification. Places are more difficult to identify than things. The plenum is apt to be a $\kappa\nu\mu\acute{\alpha}\tau\omega\nu$ $\mathring{\alpha}\nu\acute{\eta}\rho\iota\theta\mu o\nu$ $\gamma\acute{\epsilon}\lambda\alpha\sigma\mu\alpha$ [6], in which there are no stable landmarks and nothing can be definitely distinguished from anything else; and the interplay between redescription and transformation, on which we based our main argument for the Euclidean nature of space, is correspondingly less convincing.

Let us dispense with the notion of a thing, and keep only that of a person – regarding persons, however, as still being located in space, and being able to interact causally with their immediate environment. Instead of Newton's pluralism, where the prime preoccupation is to individuate the fundamental things and then to re-identify them, we have a pluralism more like that of Leibniz, where each monad is indubitably itself (every monad is qualitatively different from every other – just as people are – and each has its own quasi-personal identity), but there is a problem of relating, in a plausible fashion, any with any other. We assume a world of persons, each unique, each a centre of consciousness, each with a private memory, and capable of formulating his own plans for the future, and sometimes of carrying them out. We assume further that there is no difficulty on the part of anyone in recognizing anyone else: this, although not totally true of our own experience, is a reasonable idealization of it. We assume, finally and in view of the corporeal constitution of our friends, less plausibly, that two people may be in the same place at once, although equally they may not. For the purpose of the argument it is true enough: and the metaphysical ideal, monads, unlike atoms, do not have to be impenetrable (see § 27, pp. 126–7). Our difficulty is to give an account of how people act, or rather, interact, since there are no things, only an unstable flux. Space will be introduced not as the place where things can be, but in terms of a basic process, the process that enables people to communicate, and we shall consider the implications of their being able to communicate with one another, at a distance, in a causally coherent way. We shall show that granted certain assumptions, in particular some about parity of esteem and the featurelessness of space, the Lorentz transformations will emerge as the natural and rational way whereby widely separated observers – persons – can correlate their

[6]   Aeschylus, *Prometheus*, 89–90: "An uncountable giggle of ripples".

respective systems for labelling places and dating events. It is, so to speak, a transcendental deduction of the Lorentz transformations on the basis of a communication argument [7]. These are not the considerations which originally led to the Special Theory of Relativity, nor the ones which seem most telling or elegant to a physicist. But they are a natural analogue to the considerations that established the pre-eminence of the Euclidean group, and involve issues as much the concern of the philosopher as the physicist.

[7]   Compare similar arguments in § 8, pp. 44–7, above; and § 48, pp. 244–6, below.

§ 42
Newtonian mechanics and relativity

It is often said that relativity refuted Newton. But it is a misleading over-simplification. There is no straight opposition between relativity theories and Newtonian, absolute, theories. All geometry and all mechanics are in some degree relativistic: all concern themselves with some equivalence relations which are taken as invariant. Newton, just as much as Einstein, thinks that the laws of mechanics are origin-indifferent as regards both time and space, and orienta-tion-indifferent, as regards space; more, Newton, like Einstein, regards the laws of mechanics as being indifferent as regards uniform velocity. So far as Newtonian *physics* is concerned, we do not have to believe in absolute rest any more than in one preferred frame of reference. It is only if we are talking about that sense of philosopher's space, in which it is meaningless to speak of its either moving or not moving, or if we import extra-physical considerations from, e.g., theology, that we seem to deny the relativity that Einstein stood for (see § 30, pp. 141–2). Newton does sometimes himself so speak [1]. He confuses physical arguments against accelerating and rotating frames of reference –

[1] In Scholium to Definition VIII of the *Principia* (where he is largely concerned with philosopher's space – see above, § 30, p. 143); and in Hypothesis I of Book III, and Proposition IX. See also Samuel Clarke's Third Reply, § 4, in H. G. Alexander (ed.), *The Clarke-Leibniz Correspondence* (Manchester, 1956).

which are not criticized by the Special Theory of Relativity, but only by the General Theory – with philosophical presuppositions about philosopher's space, and with theological considerations that lie outside physics (see § 47). Philosopher's space is unmovable in the Pickwickian sense that it is neither movable nor not movable – that is, it makes no sense to speak of its either moving or not moving (see § 30. p. 142). Theologically (and not only theologically) speaking we may assign a preferred frame of reference, with a particular origin ("the centre of the system of the world" [2], or "the common centre of gravity of the earth, the sun and all the planets" [3]), which is at rest. There is no reason why we should not – God may have looked, and seen that it was so, and told Newton. Only there is no *physical* reason why we should. So far as the laws of mechanics go, any other frame of reference, not rotating and not accelerating with respect to Newton's preferred one, will do equally well, and there are no physical grounds for preferring one to any other.

The Special Theory of Relativity arose not from any inadequacies of Newtonian mechanics as such, but from problems from electromagnetism. For the speed of light (and of all electromagnetic radiation) is finite; and moreover, as revealed by the Michelson-Morley experiment, *it* is constant whatever the frame of reference with respect to which it is measured. It is the universal constancy of the speed of light that Newtonian mechanics did not accommodate. Newtonian mechanics, with its simple addition law for velocities in the same direction, would lead us to expect that the speed of light measured in two different frames of reference, moving at a uniform velocity with respect to each other, would work out differently. From which it would follow that some frames of reference would appear preferable to others, in that in them the velocity of light was the same in all directions. These we should think of as being at rest. If we have a finite speed of light and the addition rule for velocities, then we have, *eo ipso*, absolute rest. For, as we shall show shortly, every rule for compounding magnitudes – such as velocities – allows of only one "universal magnitude", which, whatever other magnitude it is compounded with, will yield the same result. If we attempt to introduce more than one "universal magnitude" we overdetermine the system, and can save its symmetry only by distinguishing one of its instances (here, the special case of absolute rest) from all the others.

The argument is a variant on those we have already adduced for the homogeneity of time (see § 14). We regraduate a one-dimensional magnitude so as to show the necessity of the conditions claimed. We deal only with the addition of velocities in the same direction, and we assume that the resultant, whatever its magnitude, must have the same direction too. We then claim that whatever our rule for compounding the magnitudes of velocities all in the same direction (though not necessarily in the same sense), it must at least satisfy the following

[2]   Hypothesis I of Book III.
[3]   Proposition XI.

six conditions. We stipulate these conditions for the compounding of magnitudes – in this case velocities all in the same direction (though not necessarily the same sense):

(i) $f(u,v) = f(v,u)$.

(ii) $f(u, f(v, w)) = f(v, f(u, w))$.

(iii) $f(u, 0) = u$.

(iv) $f(u, v)$ can be idealized as continuous and differentiable in u and v.

(v) There is an open interval of values of u and v, with $u = 0$ and $v = 0$ lying within the closure of that interval for which the derivatives of $f(u,v)$ with respect to u and v are positive and continuous.

And in those cases where $f(u,v)$ is defined for negative as well as non-negative values of u and v:

(vi) $f(-u,-v) = -f(u,v)$.

The first three of these conditions are simple and obvious.

(i) is the commutative law; if we compound one quantity with another, the result will be the same as if we had compounded the other with the one.

(ii) is the associative law, saying that it does not matter in which order we combine quantities; as, for example, $A + (B + C) = (A + B) + C$ or $A + (B + C) = B + (A + C)$. There are various forms of the associative law, equivalent to each other. We have selected the form that requires the fewest steps in the subsequent proof.

(iii) points the existence of a null or zero quantity, which added to any quantity leaves it unchanged.

Condition (iv) rules out applications to discrete quantities – e.g. numbers of people – and restricts us to the continuously varying quantities where the concepts of mathematical analysis are applicable – e.g. velocity, energy, mass.

Condition (v) is really saying that $f(u,v)$ is strictly monotonic increasing in both u and v, and generally well behaved, for all normal values of u and v. For any notion resembling that of adding it is at least necessary that if either of the quantities being combined is increased, the other remaining the same, then the resultant total should be increased also, even if not by the same amount. The phraseology is cumbrous in order to allow for the special case where a universal magnitude is combined with some other one, in which case the resultant will not be increased even though the other component magnitude is.

Condition (vi) enables us to handle negative quantities. It says that negative quantities are to be combined in the same way as positive quantities, as regards their absolute magnitudes, but keeping their negative sign. An alternative formulation of condition (vi) is

(vi') $f(u, -u) = 0$.

It is easily seen that (vi′) follows from (vi), and slightly less easily than (vi) follows from (vi′).

Conditions (ii), (iii) and (vi′) are the conditions for the quantities forming a *group* with $f(u,v)$ as the group multiplication (see § 30, p. 147). Conditions (i), (ii), (iii) and (vi) are the conditions for an Abelian group. Conditions [(i)], (ii), (iii), (iv) and (vi′) are the conditions for a [Abelian] continuous group (to be exact, we need only continuity in (iv), not differentiability). We are now showing that the addition of the monotonicity condition (v) means that the group has also a universal element, with a property analogous to that of zero in multiplication, viz. $0 \times x = 0$ for all x. It is noteworthy that if we have conditions (v) (monotonicity) and (iv) (continuity), together with (ii) (associativity), we do not need to have (i) (commutativity) in order to secure that the group be Abelian [4]. We now prove the theorem.

Differentiating condition (ii) with respect to v, considering v and w as variables, and u as fixed,

$$\left(\frac{df[u, f(v, w)]}{df(v, w)}\right) \times \left(\frac{\partial f(v, w)}{\partial v}\right) = \frac{\partial f[v, f(u, w)]}{\partial v}. \tag{1}$$

Furthermore from condition (iii)

$$\left[\frac{df[u, f(v, w)]}{df(v, w)}\right]_{v=0} = \frac{df(u, w)}{dw}. \tag{2}$$

∴ when $v = 0$, (1) becomes

$$\left(\frac{df(u, w)}{dw}\right) \times \left[\frac{\partial f(v, w)}{\partial v}\right]_{v=0} = \left[\frac{\partial f[v, f(u, w)]}{\partial v}\right]_{v=0}. \tag{3}$$

Let $\phi(t) = _{\text{Df}} \left[\frac{\partial f(v, t)}{\partial v}\right]_{v=0}$.

We can then rewrite (3):

$$\left(\frac{df(u, w)}{dw}\right) \times \phi(w) = \phi f[(u, w)]. \tag{4}$$

Now $\phi(t)$ is continuous and positive for some range of values including $t = 0$, by condition (v), ∴ the definite integral exists

$$\int_0^w \frac{dt}{\phi(t)}$$

and will be a function of w, which we shall write as $\Omega(w)$, i.e.

$$\Omega(w) = _{\text{Df}} \int_0^w \frac{dt}{\phi(t)}. \tag{5}$$

[4] See J. R. Lucas, *The Concept of Probability* (Oxford, 1970), ch. 3, pp. 34–8, for details of the alternative sets of conditions that will suffice.

We note that

$$\Omega(0) = \int_0^0 \frac{dt}{\phi(t)} = 0. \tag{6}$$

We note also that since, by condition (vi)

$$\left[\frac{\partial f(v,-t)}{\partial v}\right]_{v=0} = \left[\frac{\partial f(-y,-t)}{\partial(-y)}\right]_{-y=0} = \left[\frac{-\partial f(y,t)}{-\partial y}\right]_{y=0},$$

$\phi(-t) = \phi(t)$, so that

$$\Omega(-w) = -\Omega(w). \tag{7}$$

From (5) we obtain

$$\frac{d}{dw}\left(\Omega[f(u,w)] - \Omega(w)\right) = \left(\frac{d\Omega[f(u,w)]}{df(u,w)}\right) \times \left(\frac{df(u,w)}{dw}\right) - \frac{d\Omega(w)}{dw}$$

$$= \left(\frac{1}{\phi[f(u,w)]}\right) \times \left(\frac{df(u,w)}{dw}\right) - \frac{1}{\phi(w)}$$

$$= 0 \quad \text{from (4).} \tag{8}$$

Hence $\Omega[f(u,w)] - \Omega(w) = C$, where C is a constant independent of w. Consider the case when $w = 0$; then

$$\Omega[f(u,0)] - \Omega(0) = C$$
$$\therefore C = \Omega(u)$$
$$\therefore \Omega[f(u,w)] - \Omega(w) = \Omega(u)$$
$$\therefore \Omega[f(u,w)] = \Omega(u) + \Omega(w). \tag{9}$$

We have thus proved that for any two-argument function $f(u,v)$ that satisfies the five conditions stated on p. 199 above (which we should certainly demand if it was to be an addition rule) there is another single-valued function $\Omega(u)$, such that

$$\Omega(f(u,v)) = \Omega(u) + \Omega(v).$$

This theorem is of great importance to any theory of measurement in which only one magnitude is being measured. It means in effect that we can always regraduate our scales so that more complicated addition rules are replaced by the rule of simple addition. There is essentially only one way of "adding" quantities, though there are as many variations on it as we please.

The application of the theorem to the problem of universal speeds is clear. A universal speed, c, is defined as one such that

$$f(c,u) = c \quad \text{for all } u. \tag{10}$$

Hence, by the theorem

$$\Omega(f(c,u)) = \Omega(c) + \Omega(u) = \Omega(c) \quad \text{for all } u. \tag{11}$$

Now $\Omega(u)$ is strictly monotonic increasing over the range referred to in condition (v). These are values of u for which $\Omega(u)$ cannot be zero. But there are no finite values of $\Omega(c)$ that will satisfy (10) when $\Omega(u) \neq 0$. The only possible solutions for $\Omega(c)$ are $\Omega(c) = \pm\infty$.

The only possible values of c for which (10) will hold are the limits, if there are any, of the open interval referred to in condition (v), at which the derivative $[df(v,u)/dv]_{v=0}$ becomes itself zero, or else, if there are no limits, the values $\pm\infty$ itself. In these two cases alone the integral defined in (5) can become infinite, and condition (11) can become satisfiable [5]. Since, by (7), $\Omega(-w) = -\Omega(w)$, the two solutions of (10) differ only in sign, and can be written

$$c = \pm\, \Omega^{-1}(\infty).$$

This proves that there can be one, but not more than one, universal speed. (In any case, if there were two – say, c and c' – we should have that $(c, c') = c$, and $(c, c') = (c', c) = c'$, so that $c = c'$.) [6] We cannot hope to have a theory in which both the speed of light and an infinite speed could be combined, both being universal [7].

Newtonian mechanics has addition as its composition rule for speeds, and therefore ∞ as its universal magnitude. It thus cannot accommodate the speed of light being a universal magnitude, and only finite. It thus makes out time to be more markedly distinct from space than the Special Theory of Relativity does. For with a finite speed of light we can compare spatial intervals with temporal ones – we can measure distances in light years as well as in miles. But with an infinite speed there is no conversion ratio, and we could not trade in any one spatial distance rather than any other against a given interval of time. Time and space would be quite disparate. And, arguing the other way, since Newtonian metaphysics accords a different logical status to time and space, making the former noun-like, and the latter a glorified adjective, it needs time and space to be altogether incommensurable, and the only universal speed to be infinite. If, however, time and space are to be fused together in space-time, they need to be commensurable, and then there must be a conversion ratio that is finite. The price is to forgo absolute simultaneity – as is only to be expected; if we can "cost" space against time, we cannot expect

[5]   In the first of these two cases we can be sure that the integral will become infinite. In the second it may, but we cannot be sure that it will. For example,

$$\int_0^\infty \frac{dy}{(y+1)^2}$$

has the value 1. In such a case condition (ii) would be unsatisfiable for every c, and there would be no universal element.

[6]   I am indebted to Mr W. Newton-Smith, Fellow of Balliol College, Oxford, for drawing my attention to this simplification.

[7]   For an elegant derivation of the actual rule for the composition of velocities, granted certain other assumptions, see Sir Edmund Whittaker, *From Euclid to Eddington* (New York, 1958), § 21, pp. 49–51.

different places in space to have necessarily the same time. And, of course, we can achieve only a half success. Time will not be completely like space. The universal velocity will be represented geometrically as an isotropic line – a line whose slope is i. Even though commensurable, there will be an exchange factor of the square root of minus one.

It may still be unclear why time should not be in like case with any one dimension of space in respect of the others. We do not need isotropic lines in order to compare distances to the north with distances to the east: why should we in order to compare a temporal interval with either? In saying that space is Euclidean we have claimed that an orthogonal redescription (i.e. a reorientation) anywhere leaves geometrical properties – and therefore length in particular – unchanged; and there is a reorientation from north to east which will make a northwards length equivalent to an eastwards one. The analogue to a reorientation is a velocity. But whereas not every northward distance can be reoriented to be congruent with a given eastward distance, any distance can, by a suitable velocity, be covered in any temporal interval. Although angles and reorientations are, like velocities, continuous, they differ from velocities in that the range of possible transformations of a given interval does not cover the whole of space. By increasing the velocity we can make a given temporal interval correspond (within wide limits) to any spatial distance; but by increasing the angle we cannot make a given spatial distance be congruent with *any* other. Indeed, if we go on increasing the angle we shall only come back to where we started from. Angles are periodic; but velocities – pseudo-angles – cannot be; not, at least, without recasting a large part of our conceptual structure. Reorientations, therefore, can provide the fine-structure of a useful equivalence relation, whereas velocities constitute too wide-ranging a correlation to do so. Therefore, since velocities are not periodic or sufficiently limited in any other way, we cannot use them all as a transformation group to transform temporal intervals into spatial distances and vice versa. If there is to be a one–one correspondence between spatial distances and temporal intervals it is going to constitute something very like a unique velocity; which, if it is to be equally valid for all Galilean frames of reference, will need to be a universal magnitude, and hence, in geometric terminology, an isotropic line, and hence bound up with imaginary numbers.

§ 43
The Lorentz group of transformations

If we follow out the programme of making time as space-like as possible, we shall be led to the Special Theory of Relativity. The argument is not watertight, but it is illuminating. In this section I shall argue the analogy between the Lorentz group of transformations, which are characteristic of the Special Theory of Relativity, and the orthogonal group whose Euclidean merits in Newtonian space I have already extolled; and in the next section I shall offer an entirely different, transcendental, justification of the Lorentz transformations as being the best of all possible ways in which bewindowed but distant Leibnizian monads could re-establish a harmony between their discordant themes.

We have shown that if time is to be regarded along with space as part of space-time, there must be some conversion factor, identified with the speed of light, and conventionally known as c; and that there is still a factor of i, $\sqrt{(-1)}$, distinguishing time-like dimensions from space-like ones. In order to make the analogy between time and space as strong as possible, we need to consider not t, the measure of temporal durations, but ict, which we shall write x_4. We write the other, space-like, coordinates as x_1, x_2, x_3. Any transformation in three dimensions must have at least one eigenvector (since a cubic equation always has a real root), and therefore an orthogonal transformation can be represented, with suitable choice of axes, by the matrix

$$\begin{bmatrix} \cos\theta & \sin\theta & 0 \\ -\sin\theta & \cos\theta & 0 \\ 0 & 0 & \pm1 \end{bmatrix}$$

where θ is a parameter, and in fact represents the angle of each particular reorientation. We can discard the minus sign as corresponding to an improper reorientation which, since it would involve a breach of continuity, could not correspond to any physical process.

If we generalize to four dimensions, while still remembering that the four dimensions are still not quite *pari passu*, we shall have as the analogue of an orthogonal transformation

$$\begin{bmatrix} \cos\phi & 0 & 0 & \sin\phi \\ 0 & 1 & 0 & 0 \\ 0 & 0 & 1 & 0 \\ -\sin\phi & 0 & 0 & \cos\phi \end{bmatrix} \quad \text{or} \quad \begin{bmatrix} 1 & 0 & 0 & 0 \\ 0 & 1 & 0 & 0 \\ 0 & 0 & \cos\phi & \sin\phi \\ 0 & 0 & -\sin\phi & \cos\phi \end{bmatrix}.$$

Leaving out the directions in which there is no change, and rewriting ϕ as $i\psi$ [1], the matrix becomes

$$\begin{bmatrix} \cosh\psi & i\sinh\psi \\ -i\sinh\psi & \cosh\psi \end{bmatrix}$$

where ψ is a parameter, the "pseudo-angle" between the two sets of axes. In this comparison, a pseudo-angle should correspond to a uniform velocity, and be compared with a change of orientation. And just as a change of orientation was a matter of indifference, so a uniform velocity is to be. So too accelerations and angular velocities are seen to be in like case, and both, in Newtonian mechanics, involving a "force".

The comparison of time with a dimension of space, to within a factor of ic, is important and puzzling. It shows to what extent time can, and cannot, be regarded as analogous to space. The factor ic does not affect the topological property of continuity, but does have a bearing both on metrical properties and on such topological properties as whether one could have a closed curve in space-time in the same way as one can in space. Minkowski geometry is markedly different from Euclidean geometry – as we have noted, a line in Minkowski geometry can be at right angles with itself (§ 32, p. 158). Moreover the (3+1)-dimensional geometry of space-time is very different from a (2+2)-dimensional geometry (two space-like and two time-like). It is therefore dangerously misleading to speak as though time were just a fourth dimension of space. It is *a* dimension of space-time, but the odd man out, and different from the other three.

[1]   It is easily verified that $\tan i\psi = i\tanh\psi$, $\sin i\psi = i\sinh\psi$, $\cos i\psi = \cosh\psi$.

The need for the pure imaginary factor ic parallels the part played by periodicity in our experience of time. As we have seen, time seems naturally, if not absolutely necessarily, periodic in a way that space does not, and we come across innumerable systems whose temporal development is periodic or almost periodic (§ 14, p. 83; § 12, p. 70). All continuous periodic functions can be expressed by means of a series of trigonometrical functions ($\cos x$, $\sin x$, etc.) and these can be expressed as the exponential functions of imaginary numbers (e.g. $\cos x = \frac{1}{2}(e^{ix} + e^{-ix})$). Whereas the exponential function of a real number is monotonically increasing, that of an imaginary number is periodic, in much the same way as our untutored intuition of space is of pure unbounded extension going on indefinitely without limit, whereas periodicity seems characteristic of time. It is perhaps only a mathematical accident, and it may be fanciful to connect this with the square root of minus one, the factor i, which is always associated with the time-like dimension in the theories of relativity, but I commend it as a speculation to the reader.

In order to apply the analogy, we need to translate back from x_4 to ict. The velocity measured in conventional units will need to be expressed as a fraction of c, and will then be analogous to a gradient: i.e. v/c will be $\tanh\psi$. Under this redescription the analogue of the orthogonal group becomes the Lorentz group of transformations, which we could now write as

$$x = \cosh\psi\,(x' + vt')$$
$$y = y'$$
$$z = z'$$
$$t = \cosh\psi(t' + vx'/c^2).$$

$\cosh\psi$ can be shown to be $(1 - v^2/c^2)^{-\frac{1}{2}}$ and we thus obtain the Lorentz transformations in their canonical form [2]:

$$L \text{ (i) } \quad x = \left(1 - \frac{v^2}{c^2}\right)^{-\frac{1}{2}} (x' + vt')$$
$$\text{(ii) } \quad y = y'$$
$$\text{(iii) } \quad z = z'$$
$$\text{(iv) } \quad t = \left(1 - \frac{v^2}{c^2}\right)^{-\frac{1}{2}} \left(t' + \frac{vx'}{c^2}\right)$$
$$v = -v'$$

for two frames of reference, with their axes oriented in the same directions, and moving with a uniform velocity, v, along the X-axis, with respect to each other. The simple case saves heavy, and unnecessary, algebraic working. Once we have mastered it, we can generalize the transformations to deal with cases where the axes are not oriented in the same direction, or the relative velocity is not in the direction of the X-axis of either frame of reference.

[2]   The reason for printing in two colours will emerge in the next few pages.

The Lorentz transformations are not mathematically very formidable, but are difficult to grasp conceptually. They are best understood as a set of *translation rules*, enabling us to translate from one mathematical language to another, which, although very similar, differ in the way they refer to places and dates. The Lorentz transformations are a set of rules for translating from one frame of reference to another, one coordinate system to another. It is like having a rule for translating A.U.C. (*Ab Urbe Condita*) into A.D./B.C. (*Anno Domini*/Before Christ), or Olympiads into A.D./B.C., or grid references in the National Grid of the United Kingdom into latitude and longitude.

We consider two persons, Red and Blue, who are in communication with each other, and work out what rules they must adopt if each is to translate the other's formulae and references into his own, granted certain assumptions of symmetry, uniformity and non-egocentricity. For ease of understanding, all Red's formulae, references and arguments will be printed in red, and all Blue's formulae, references and arguments in blue. Blue's formulae and references will also carry dashes or primes, e.g. x', to make it easier to read aloud, and also to correspond to the more normal, monochromatic, usage of physicists. We can obtain further clarity by thinking of Red and Blue as speaking different tongues, e.g., of Red speaking Russian, say, and Blue Greek, although strictly speaking their languages differ only in their noun vocabulary, the way in which they refer to places and dates, and not in their syntax, nor in many of their verbs and adjectives.

We also need to make it clear not only *from* what language *into* what language the translation is being made, but *by whom* it is being made. We have not only Russian–Greek and Greek–Russian dictionaries, but dictionaries published in Russia and dictionaries published in Greece, and often there are significant differences between them. We need to distinguish between Red's rules for translating his terms into Blue's terms or vice versa, and Blue's rules for translating his terms into Red's terms or vice versa. I mark this by the colour of the 'equals'. A red 'equals' means that it is Red's rule, a blue 'equals' means that it is Blue's rule. Thus the transformation

$$L \quad \text{(i)} \quad x = \left(1 - \frac{v^2}{c^2}\right)^{-\frac{1}{2}}(x' + vt')$$

$$\text{(ii)} \quad y = y'$$

$$\text{(iii)} \quad z = z'$$

$$\text{(iv)} \quad t = \left(1 - \frac{v^2}{c^2}\right)^{-\frac{1}{2}}\left(t' + \frac{vx'}{c^2}\right)$$

is Red's rule for translating his red terms x, y, z, t into Blue's terms x', y', z', t'. It is a Russian–Greek dictionary published in Russia.

We have four sets of transformation rules to consider: Red's rules for translating red into blue, Red's rules for translating blue into red, Blue's rules for translating red into blue, and Blue's rules for translating blue into red.

Schematically, we have [3]:

	Red's rules	Blue's rules
Red into blue	$\{x, y, z, t\} = L\{x', y', z', t'\}$	$\{x, y, z, t\} = M\{x', y', z', t'\}$
Blue into red	$\{x', y', z', t'\} = N\{x, y, z, t\}$	$\{x', y', z', t'\} = \Lambda\{x, y, z, t\}$

Each of $LMN\Lambda$ (L, N are capital Roman letters, M, Λ are capital Greek letters) is a whole set of rules, a whole dictionary. We have given L (p. 207) and shall give Λ. We do not normally need to give M and N, because M is the inverse of Λ and N is the inverse of L. If Red has adopted a set of rules L for translating red into blue, then he has implicitly adopted a certain set for translating blue into red. If a Russian has got a Russian–Greek dictionary already, then he would land himself in an inconsistency if he went on to adopt highly eccentric rules for translating Greek back again into Russian. Consistency demands that if he translates a piece of Russian into Greek and then back again into Russian, the result shall be recognizably the same as the original. Therefore if Red goes

$$\{x, y, z, t\} \xrightarrow{L} \{x', y', z', t'\} \xrightarrow{N} \{x, y, z, t\}$$

the final result must be the same as the original. That is, transformation L followed by transformation N is an identity transformation, $NL = I$, and N is the *inverse* transformation of L and vice versa. We write this

$$N = L^{-1}.$$

A similar argument applies to Blue's rules, and shows that transformation Λ followed by transformation M is an identity transformation, i.e.

$$MA = I$$
$$M = \Lambda^{-1}.$$

At the risk of wearisome repetition I give all four rules in full, and comment on the differences between them. The two points to be made are that whereas L and N (and Λ and M) are *inverses*, L and Λ (and M and N) are *converses*, and that this is associated with the fact that $v = -v'$ and vice versa. For if the X-axes of the two frames of reference are oriented in the *same* direction, then the velocities of the one frame of reference with respect to the other frame of reference will be in opposite directions in the two cases. If you are due west of me and we each have a relative velocity towards the other of 4 m.p.h. then I shall be going 4 m.p.h. *west* with respect to you, and you will be going 4 m.p.h *east* with respect to me. Hence we need the minus sign before the v' on the right hand side.

[3]   It is easy to be confused about the direction in which the translation goes. L is a set of rules expressing Russian referring expressions (x, y, z, t) in terms of Greek ones (x', y', z', t'), and so is appropriate for replacing (x, y, z, t) in a *Russian* text by their Greek equivalent: i.e. translating from Russian into Greek.

We have already given L

$$L \ \text{(i)} \quad x = \left(1 - \frac{v^2}{c^2}\right)^{-\frac{1}{2}}(x' + vt')$$

(ii) $\quad y = y'$

(iii) $\quad z = z'$

$$\text{(iv)} \quad t = \left(1 - \frac{v^2}{c^2}\right)^{-\frac{1}{2}}\left(t' + \frac{vx'}{c^2}\right)$$

and $v = -v'$

From these the inverse transformation N, Red's rules for translating blue into red, follows by simple algebra. From L(i) we have

$$(x' + vt') = \left(1 - \frac{v^2}{c^2}\right)^{+\frac{1}{2}} x$$

and from L(iv), multiplied by v,

$$\left(vt' + \frac{v^2 x'}{c^2}\right) = \left(1 - \frac{v^2}{c^2}\right)^{+\frac{1}{2}} vt$$

$\therefore$ subtracting $\qquad x' - \dfrac{v^2 x'}{c^2} = \left(1 - \dfrac{v^2}{c^2}\right)^{+\frac{1}{2}}(x - vt)$

$$\left(1 - \frac{v^2}{c^2}\right)x' = \left(1 - \frac{v^2}{c^2}\right)^{+\frac{1}{2}}(x - vt)$$

$$x' = \left(1 - \frac{v^2}{c^2}\right)^{-\frac{1}{2}}(x - vt)$$

which is N(i).

From L(ii), $\quad y = y' \qquad y' = y$ which is N(ii).

From L (iii), $z = z' \qquad z' = z$ which is N(iii).

From L(i), multiplied by v/c^2,

$$\frac{v}{c^2} x' + \frac{v^2}{c^2} t' = \left(1 - \frac{v^2}{c^2}\right)^{+\frac{1}{2}} \frac{vx}{c^2}$$

From L(iv)

$$t' + \frac{v}{c^2} x' = \left(1 - \frac{v^2}{c^2}\right)^{+\frac{1}{2}} t$$

$\therefore$ subtracting the former from the latter,

$$t' - \frac{v^2}{c^2} t' = \left(1 - \frac{v^2}{c^2}\right)^{+\frac{1}{2}}\left(t - \frac{vx}{c^2}\right)$$

$$\left(1 - \frac{v^2}{c^2}\right)t' = \left(1 - \frac{v^2}{c^2}\right)^{+\frac{1}{2}}\left(t - \frac{vx}{c^2}\right)$$

$$t' = \left(1 - \frac{v^2}{c^2}\right)^{-\frac{1}{2}}\left(t - \frac{vx}{c^2}\right)$$

which is N(iv).

We thus have established N, Red's rules for translating blue into red. Λ, Blue's rules for translating blue into red, must be essentially the same. Although there is a distinction between Greek–Russian dictionaries published in Russian and Greek–Russian dictionaries published in Greece, they cannot differ very much, or they could not both be described as being Greek–Russian dictionaries. N and Λ are both sets of rules for transforming (x', y', z', t') into (x, y, z, t); they must be essentially the same. The only difference is that whereas N is formulated in terms of v, Blue's velocity, i.e. the velocity ascribed by Red to Blue, or in Russian, скорость Синего, Λ is formulated in terms of v', Red's velocity, i.e. the velocity ascribed by Blue to Red, or in Greek, $\tau\hat{\eta}s\ \tauο\hat{v}\ \Pi\upsilon\rho\rhoο\hat{v}\ \tau\alpha\chi\upsilon\tau\acute{\eta}\tauοs$. But $v = -v'$. We have not as yet given a complete justification of this assumption, although it is clearly plausible, in view of the terrestrial example above. If the assumption be granted, we see that $v^2 = v'^2$; we also have, by the argument of the last section, that c is the same in all frames of reference. From this it follows that

$$\left(1 - \frac{v^2}{c^2}\right)^{-\frac{1}{2}} = \left(1 - \frac{v'^2}{c^2}\right)^{-\frac{1}{2}}.$$

We therefore shall be able to abbreviate it to β, since it occurs so often, and is the same in both frames of reference. We note that when $v = 0$, $\beta = 1$; when $0 < |v| < c$, $\beta > 1$; as $|v| \to c$, $\beta \to \infty$; also that $\beta = \cosh\psi$ on p. 206 above. Λ will be the same as N in respect of β, and will differ from it only in needing to be expressed in terms of v' instead of v. We therefore write down the following equations for Λ:

Λ (i) $x' = \beta(x + v't)$

(ii) $y' = y$

(iii) $z' = z$

(iv) $t' = \beta\left(t + \dfrac{v'x}{c^2}\right)$

$v' = -v$

We can obtain M from L in an exactly similar fashion, replacing the $+v$ of L by $-v'$ in M. The reader may like to obtain M also from Λ in the same way as we derived N from L. It is a useful exercise for those unused to handling the Lorentz transformations, although excessively tedious to those already familiar with them.

§ 44
The transcendental derivation of the Lorentz transformations

There are many different ways of deriving the Lorentz transformations. The physicist may be concerned to unite the theories of electricity and of magnetism into one harmonious whole, the mathematician to find an invariant form for ds^2. Historically it was consideration of the implications of the Michelson-Morley experiment that carried most weight with scientists. It is, however, possible to argue on a much more slender basis, and to show that the same considerations we adduced at the end of the last section to secure a conceptual grasp of what the Lorentz transformations really meant are enough also to show that they must take the form that they actually do. The argument is due to Whitrow and Milne [1]. It is a communication argument (see above § 8, pp. 44–7; and below § 48, pp. 244–6). We consider what rules Red and Blue must adopt if, being in communication with one another, each is to translate

[1]   G. J. Whitrow, *The Natural Philosophy of Time* (Edinburgh, 1961), ch. III, § 8, pp. 171–3, and ch. IV, §§ 3–4; E. A. Milne, *Modern Cosmology and the Christian Idea of God* (Oxford, 1952), ch. III and IV; and *Kinematic Relativity* (Oxford, 1948), ch. II, esp. § 24. The key regraduating theorem was proved by G. J. Whitrow, *Quarterly Journal of Mathematics*, VI (1935), para. 4, pp. 252–6. See also E. A. Milne and G. J. Whitrow, *Zeitschrift für Astrophysik*, XV (1938), pp. 273–4; A. G. Walker, *Quarterly Journal of Mathematics*, XVII (1946), pp. 67–8; W. H. McCrea, *Proceedings of the Royal Irish Academy*, XLV (1938), A, pp. 24–5.

the other's references into his own system, subject to certain conditions of uniformity and parity of esteem between observers. It plays the same part in the development of the Leibnizian space-time as impenetrability did in that of Newtonian space, but of course will be less attractive to the physicist than to the philosopher.

Communication cannot be instantaneous in space-time. For then we should have to ascribe to it an infinite speed – which would be a universal speed, and the only universal speed (unless we abandoned parity of esteem between observers, and accepted some frames of reference as being at absolute rest and others as being in absolute motion). But we must have a finite universal speed if space and time are to be commensurable and capable of being taken together; and hence no infinite speeds whatever.

We must also demand that the speed of communication be finite, if we are to be able to give any causal account of communication. Telepathy may be like Newtonian gravitation, instantaneous in its effect, but then it becomes an occult phenomenon. Causal causes, as we claimed earlier (§ 8, pp. 50–1), are necessarily earlier than their effects, and therefore in any causal theory of communication the effect – the reception of a message – must occur after the cause – the sending out of the signal. If we add the further requirement of continuity (§ 41, p. 194), the time taken for the message to be conveyed will have to be a strictly monotonic increasing function of the distance covered. We therefore again will require a finite speed of communication.

Neither of these arguments shows that the speed of light is a universal speed; or indeed that communication is propagated with the velocity of light. Red and Blue are in communication, as it were by wireless, but it need not be by radio waves or light signals; all we need is that they are using the *best* means of communication they can. It is true, but only contingently true, that the best method of communication we think there is is by electromagnetic radiation – radio, radar or light.

Let each message contain the information that it was dispatched (or "transmitted") at such and such a time according to the time reckoning of the person who dispatched it; furthermore, when a message is received, let an acknowledgement be at once dispatched in return. The acknowledgement will say both when it itself was dispatched – according to the time reckoning of the person who dispatched *it* – *and* what message it is answering. It will be the equivalent of the businessman's letter:

3 November 1964   Birmingham

Dear Sir,
    Thank you for your letter of the 2nd inst., which is receiving attention.

Yours faithfully,
A

Suppose Red and Blue exchange a series of messages thus:

t_1 Home (0, 0, 0)

Dear Blue,

   I hope this finds you as it leaves me.

                               Yours ever, Red

Or, in Russian:

t_1 Дома (0, 0, 0)

Дорогой Синий,

   Надеюсь, что при получении зтого письма вы будете в том-же хорошем состоянии здоровья, как я теперь.

                               С приветом, Красный

To which Blue replies:

t_2' Home (0', 0', 0')

Dear Red,

   Thank you for your letter, dated 't_1', which has just arrived.

                               Yours ever, Blue

Or, in Greek:

t_2' Οἴκοθεν (0', 0', 0')

Κυάνεος Πυρρῷ χαίρειν,

   ἥσθην ἐπιστολὴν ἀρτίως δεξάμενος ᾗ ὥραν 't_1' ἐπέγραψας.

Red then answers Blue's letter:

t_3 Home (0, 0, 0)

Dear Blue,

   Thank you for your letter, dated 't_2'', which has just arrived.

                               Yours ever, Red

t_3 Дóма (0, 0, 0)

Дорогой Синий,

   Спасибо за ваше письмо с датои «t_2'», которое я только что получил.

                               С приветом, Красный

To which Blue again replies:

t_4' Home (0', 0', 0')

Dear Red,

   Thank you for your letter, dated 't_3', which has just arrived.

                               Yours ever, Blue

$$t_4'$$ $$Oἴκοθεν\ (0', 0', 0')$$

Κυάνεος Πυρρῷ χαίρειν,
ἤσθην ἐπιστολὴν ἀρτίως δεξάμενος ᾗ ὥραν ʻt_3ʼ ἐπέγραψας.

From this exchange of messages, Red knows when he sent his first message, when he received one back – когда он послал первое письмо, и когда он получил ответ and also knows the date on which Blue says he, Blue, received the message according to his, Blue's, reckoning – αὐτός κατά γε τὸν αὐτοῦ λογισμον, τὴν ἀγγελίαν δέξασθαι. Similarly Blue knows when he sent his first answer, when he received an answer acknowledging it – ὁπηνίκα τε τὴν πρώτην ἀπόκρισιν ἀπέστειλε, καὶ ὁπηνίκα τὴν ταύτης ἀπόκρισιν ἐδέξατο and also knows the date on which Red says he, Red, received the message according to his, Red's, reckoning – когда он, Красный, по своему расчёту получил письмо.

But neither Red nor Blue are prepared to accept the other's *ipse dixit* as conclusive evidence for dates. Quite often I do not. If I write abroad to a philosopher in January, and in June receive an answer dated February, I am often a bit sceptical – it depends on the man, and on what I know of his country's postal system. I do not always and necessarily accept that a letter headed February was actually written in February, as we know it in Britain. I do not need to impute dishonesty to a colleague who *says* that he is writing it in February. It is merely that he is writing from another country, and they do things differently in other countries. But not entirely differently. We have seen already that there are certain topological properties of time, which are invariant from observer to observer. I would not accept it if a man – even a philosophical colleague – dated his answer to my second letter before his answer to my first. However his time reckoning is related to mine, it must be such that a later time on mine will correspond to a later time on his, and vice versa; his time reckoning, considered as a function of my time reckoning, must be strictly monotonic increasing. But, subject to this one proviso, I am prepared to accept a colleague's word for it that he is truthfully reporting the date according to his own country's dating system, which differs from country to country.

Nevertheless, I cannot accept another man's dating system as my own. I live in Britain and go by British time, and I want to know when, by British time, the man received and answered my letter. Now what can I say? I know that he cannot have received my letter, by British time, before I sent it; nor could he have answered it after I had received his answer. If I sent the letter in January and received an answer in June, it must have reached him some time between January and June inclusive. Any date between that of dispatch and that of receipt of the answer is possible – at least, will not lead to a contradiction. We can express this in general terms by saying

$$t_2 = \epsilon t_1 + (1 - \epsilon)\, t_3$$

where t_2 is the date Red is to *ascribe* in his (Red's) time reckoning to the event constituted by Blue's receiving Red's first message, and where ϵ is a constant between 0 and 1 inclusive. If I had a very high regard for my own country's postal system I might put ϵ equal, or very nearly equal, to 1. That is, I would think that letters arrived as soon as, or almost as soon as, they were sent, and that all the delay was in getting the answer back. If I took a more realistic view of my own country's efficiency, I would set a lower value on ϵ, and think that answers came promptly, and all the delays were on the sending side.

In fact we set $\epsilon = \frac{1}{2}$. The rule then becomes

$$
\begin{aligned}
t_2 &= \tfrac{1}{2} t_1 + (1 - \tfrac{1}{2}) t_3 \\
&= \tfrac{1}{2} t_1 + \tfrac{1}{2} t_3 \\
&= \tfrac{1}{2} (t_1 + t_3).
\end{aligned}
$$

That is, the date of the arrival of the message is assumed to be exactly halfway between the date of its dispatch and the date of the receipt of its acknowledgement: it is the arithmetic mean of the two dates between which we know that it must lie.

There is a good deal of confusion about this assumption, which is a natural one, but not a necessary one. It is sometimes said that it is a *convention* that we merely *agree* to ascribe the date $\frac{1}{2}(t_1 + t_3)$ to distant events. But this suggests that we could *just as well* have ascribed some other date, in accordance with some other, conventional, value for ϵ; the point of having a convention is where we have got two or more alternatives, with nothing to choose between them in themselves, but it is important we should all choose the same one; where it does not matter what we do, provided we all do the same thing – e.g. the rule of the road, spelling, brackets. The assumption about the date of distant events is not a convention in that sense. Although the value $\epsilon = \frac{1}{2}$ is not forced on us, in the sense that any other value would lead either to a contradiction or a prediction that would turn out to be false, it is the one we should go for in the absence of any reason for any other one. When I was talking of other values I might accept, I had to say – in order to make the story plausible – that my own country's postal system was exceptionally good or exceptionally bad, or that I had a low opinion of those of other countries. When Aristotle puts forward his doctrine of the mean, he has to warn us against the presumption that the mean is an arithmetic mean [2] – which shows that this would be a natural assumption to make in the absence of any further considerations.

With the communication argument, not only is there no reason for not putting $\epsilon = \frac{1}{2}$ but there are some reasons for doing just this. We could, as a first step, invoke the principle of *non-egocentricity*. The view that "my country is better than yours" and has a "better postal system" contravenes this principle. So too, does the principle of anti-patriotism: "my country is less good than

[2] *Nicomachean Ethics*, II, 6, 1106a 29–1106b 7.

yours". Red is not entitled, according to this principle, to claim My messages travel faster than yours do, Blue – Мои письма идут быстрее ваших, Синий.

The principle of non-egocentricity needs to be invoked somewhat sparingly, or rather, very carefully and precisely; for although Red is trying not to be egocentric in the sentiments he utters about Blue's messages, he is uttering these sentiments in his own language, Russian, and is in some respects necessarily adopting his own egocentric point of view. In particular, he is assuming that he is not moving. Obviously, if one assumes that one is moving, one will expect the return journey to take longer or shorter to reach one. This is why we go out to meet the bearers of good tidings. But in setting $\epsilon = \frac{1}{2}$, we assume that every observer does not consider the possibility that he might move. It might be thought to be a serious and unacceptable contravention of the principle of non-egocentricity. It is more, however, a defect of this mode of presentation. For what we are really concerned with is not the observer but the frame of reference. The observer only represents *the origin of* the frame of reference. And it is logically impossible for the origin of a frame of reference to move with respect to that frame of reference; and so it is conceptually impossible for an observer to consider the possibility of moving.

There are two strands of egocentricity in our account of time and space, an egocentricity of reference and an egocentricity of knowledge. Because time and space are featureless, we have to impose some coordinate system as a frame of reference, and our choice is arbitrary, and the system I actually use is in use because I decided to use it. Red similarly must use some coordinate system, although not any particular one, and the one he uses will be his by his *fiat*, and to that extent egocentric, and ineliminably so. Our human knowledge of time and space is also egocentric, because it is based on experience, the experience of some one, some observer, some subject, who has his own particular point of view. Although it is inconceivable that anyone could set up a frame of temporal reference in which he could not in principle refer to his own dates, it does not seem absolutely necessary that our knowledge of space should be thus conditioned (see above § 9, p. 58, and further below § 52, p. 280). But our approach to space-time is through and through epistemological. Red and Blue are centres of consciousness, each with his own point of view, each constructing out of his own experience a temporal and spatial framework for referring to distant events – events, that is, that are distant from him. Although Red and Blue are trying to correlate their varying egocentric viewpoints as non-egocentrically as possible, they could not achieve complete non-egocentricity, because then neither would be conscious of himself or of the other, or be able to communicate with the other, or tell the other what he, himself, was experiencing. Egocentricity cannot be completely eliminated, nor can questions of epistemology. Newton may take a God's-eye view of space, but Red, like Leibniz, asks "How do I know where Blue is, and what he is doing, and when?" and therefore he has to view distant events from where *he is*, which he calls "home" (or, in mathematicians' language, the spatial origin $(0,0,0)$); and if distant objects appear to move, each observer ascribes the movement to them, not to himself. He is – necessarily – ascribing the position and movement of events and objects *relative to himself*; and he is – necessarily – at rest, relative to himself.

Setting $\epsilon = \frac{1}{2}$ requires the assumption not only that Red is at rest, but that the speed of communication is the same in both directions. Red certainly would not be justified in supposing that outgoing messages travelled faster than incoming ones just because they were going in the opposite direction, for that would be to deny the isotropy of space, that no direction is different, with regard to speed of communication, from any other one; and, in particular, that the direction from Red to Blue is no faster or slower than the opposite direction from Blue to Red. But Red might distinguish between his messages and Blue's, not because they were travelling in different directions, nor just because Blue was a different person from himself, but because I, Red, am at rest, whereas you, Blue, are moving – я, Меподвижен, но вы, Синий, двигаетесь and arguing in justification, Moving objects return messages at different velocities from those of messages returned by stationary objects – Подвижные предметы возвращают сообщения на других скоростях, чем неподвижные предметы. It is intuitively plausible. If I throw a ball at the front of an express train, it is bounced back at a much greater speed than I threw it with. And the Doppler effect shows that the movement of the transmitting source does have *some* effect on the waves transmitted. Red might well suppose that light or radio waves were bounced back at greater or less speeds according to the speed at which the object was moving – or indeed according to the acceleration or something. But in setting $\epsilon = \frac{1}{2}$, Red assumes that this is not so, and that the speed of communication is the universal speed, which remains the same even when compounded with some other speed [3]. Is it a justifiable assumption? It is, inasmuch as Red is using the *best* – that is, the *fastest* – means of communication there is. For the universal speed will be the fastest possible one, since the regraduating function, $\Omega(w)$, is monotonically increasing [4]. We have already postulated that Red should use the best means of communication he can (p. 212), but it always remains logically possible that the best he can in practice is not the best he could in principle. To this extent, there is an empirical assumption in any actual application of the argument: that we have already reached the limit in speeding up our communications. But it is an assumption that it is reasonable to make in the absence of evidence to the contrary. Once we have made this assumption, it is reasonable to lay down that Red and Blue shall both assign the same value

[3]   For discussion of the difficulties in supposing that the velocity of light depends on the motion of the emitting or reflecting body, see W. Pauli, *Theory of Relativity*, tr. G. Field (Oxford, 1958), § 3, pp. 5–9.

[4]   § 42, pp. 200–1: $\dfrac{d\Omega(w)}{dw} = \dfrac{1}{\phi(w)}$, which is never negative.

to c. Indeed, it is the only way they can achieve a common scale for measuring distances and velocities. For they are a long way apart, and cannot lay off their individual rulers against each other. They need some convention to establish equality of distances and velocities, and the only intersubjective experience they have is the temporal ordering of certain events and the fact that they can communicate with a finite speed which each has reason to regard as the universal speed of his space-time. It is therefore necessary to measure distances by reference to the speed of light (as we do when we measure in terms of light years), and velocities by comparison with it. It is for this reason that we did not need to write c or c in the Lorentz transformations (on pp. 206, 209–10), but only c.

Granted these assumptions, an observer can set up an exact dating system for distant events. Red will say Blue must have got my message at t_2, where $t_2 = \frac{1}{2}(t_1 + t_3)$, and t_1 was the date on which I sent my message, and t_3 was the date on which I received Blue's reply – Синий должен был получить моё письмо в t_2, тде $t_2 = \frac{1}{2}(t_1 + t_3)$, и где t_1 было датой, когда я послал моё письмо, и t_3 было датой, когда я получил ответ от Синего. Similarly, Blue will say Red must have got my message at t_3', where $t_3' = \frac{1}{2}(t_2' + t_4')$, and t_2' was the date on which I sent my message, and t_4' was the date on which I received Red's acknowledgement – ἐπείπερ τὴν μὲν ἐμὴν ἀγγελίαν ὥρᾳ t_2' ἀπέστειλα, τὴν δὲ τοῦ Πυρροῦ ἀπόκρισιν ὥρᾳ t_4' ἐδεξάμην, ἔοικεν ὁ Πυρρὸς τὴν ἐμὴν ἀγγελίαν ἐκείνῃ τῇ ὥρᾳ, t_3', ᾗ τούτων μεσαιτάτη ἀριθμεῖται. This can be done not only for distant observers, where we need the slightly implausible assumption of an instantaneous acknowledgement of messages, but for any objects to which we can send pulses of electromagnetic radiation and get them *reflected* back. This is what we do in radar, and also on those occasions at night when we illuminate an object by a torch or searchlight.

We can use the radar principle not only to date distant events, but also to place them. In radar, the distance is calculated from the interval between sending out the radar pulse and its return after having been reflected. On this principle Red assigns to Blue the distance l_2 from himself, Red, where $l_2 = \frac{1}{2}c(t_3 - t_1)$, and similarly Blue assigns to Red the distance l_3' from himself, Blue, where $l_3' = \frac{1}{2}c(t_4' - t_2')$. As the subscripts indicate, these distances are assigned to objects at the (assigned) date midway between the dispatch and receipt of the radar pulse. Thus the radar pulse assigns both a date, and a corresponding distance, to a distant event or object, the date being calculated from the *sum* of the dates of dispatch and receipt, the distance being calculated from the *difference* between the dates of dispatch and receipt.

The rationale of the radar rule for distance is clear. The observer is at rest with respect to himself. The message, therefore, will have to go the same distance on the return journey as on the outward journey. If the length of the single journey is l, the total journey will be of length $2l$, and since the speed is assumed to be the same in both directions, the total time interval taken for the journey will be $2l/c$. Hence $l = \frac{1}{2}c \times$ the interval between dispatch and receipt of the message.

In assigning distances, as in assigning dates, we are assuming that there is no fundamental difference between observer and observed, that the velocity of light has the same magnitude in all directions, that the observer is at rest, and that any movement of the observed will not affect the velocity with which the reflected radar pulse will travel. If we can add, further, that radar pulses go in straight lines, we can tell not only how distant a distant event is, but in what direction it lies, and hence we can locate its position completely.

The radar rules enable an observer to assign dates and positions to any, and every, distant event he knows of. These dates and positions are based entirely on the observer's own experience; they are constructs from his own dating, within his own immediate experience, of dispatching and receiving messages. Then, granted certain assumptions, it is reasonable for, e.g., Red to assign to the event of Blue's receiving and returning Red's message the date

$$t_2 = \tfrac{1}{2}(t_1 + t_3)$$

and the distance (in a certain direction)

$$l_2 = \tfrac{1}{2}c(t_3 - t_1).$$

This is very satisfactory from the point of view of Red, but Blue may not like being redated or relocated by Red. Similarly, I who live in Britain may be quite happy to assign British dates and British positions to the letter-writing activities of my correspondents in other countries, but they may not be equally pleased. I may lump Canada, U.S.A., Mexico and Brazil together as countries in the far West, quite close to one another, though miles from here – but they may not recognize themselves as neighbours, and may regard the countries of the Old World as having very strange notions of space, just as they do in fact regard them as having very strange notions of most other things, including time. Merely to impose my own dating system on other people is not a generally satisfactory solution to the problem of reconciling our discrepant dating systems. It contravenes the first principle of non-egocentricity. It is all very well for Red to say

$$t_2 = \tfrac{1}{2}(t_1 + t_3)$$
$$l_2 = \tfrac{1}{2}c(t_3 - t_1)$$

but Blue can equally well say

$$t_3' = \tfrac{1}{2}(t_2' + t_4')$$
$$l_3' = \tfrac{1}{2}c(t_4' - t_2')$$

and there is nothing to choose between them. The radar rules enable us to date and place distant objects, but do not enable us to reconcile the discordant dates recorded by different observers.

If we are not prepared to say simply that others are wrong, then we must try and find a way of reconciling the different dates recorded by different observers. If we can remain in communication with an observer for a reasonable length of time, this is possible. If I keep writing to Baghdad and getting back answers in which the date head does not have 1964 but 1384 instead, after a bit I shall twig the rule, and be able to convert Mohammedan into Christian dates, and vice versa. Such trivial differences in dating systems are easily ironed out. Essentially they are differences either in the zero (A.D. or A.H. or A.U.C., etc.), or in the unit (day, week, year, Olympiad). These are conventions in the proper sense of the word, and can be altered easily by agreement; for example, that we should both date our correspondence A.U.C. in order not to prejudge any issues between Christianity and Islam.

Red and Blue can iron out differences in their zeros very easily. For, since they are moving away from each other with a non-zero velocity, there was a date when, if the velocity had been uniform throughout, they would have been in the same place; and this date provides a natural and symmetrical starting point of the time reckoning of each. Let them both agree to measure time from then: and let us assume that this has already been done, so that $0 = 0'$.

We cannot assume that the unit will be so easily synchronized as the zero. All we know is that messages dispatched later will arrive later, so that Blue's time reckoning for receipt of messages must be a strictly monotonic increasing function of Red's time reckoning for the dispatch of messages, and vice versa; i.e. $t_2' = f(t_1)$ and $t_3 = g(t_2')$, where $f(w)$ and $g(w)$ are strictly monotonic increasing functions of w, which correlate not only the dates t_1 and t_2', and t_2' and t_3, but any dates when a message is dispatched by one and received by the other. More carefully:

(i) If t is the date in Red's time reckoning when he dispatches a message to Blue, and t' is the date in Blue's time reckoning when he receives the message, then

$$t' = f(t) \text{ [5]}$$

and (ii) if t' is the date in Blue's time reckoning when he dispatches a message to Red, and t is the date in Red's time reckoning when he receives the message, then

$$t = g(t').$$

Given a sufficient number of messages Red and Blue can both discover what the functions $f(w)$ and $g(w)$ are, more or less. Each of them will be correlating the numbers given as dates by his own time reckoning with the numbers reported in the messages being the date of dispatch according to the time

[5]   Note that in this formal definition we have correlated the date of departure with the date of arrival. When we are using subscripts t_1, t_2, t_3, we are *not* correlating the dates with the same subscripts according to this principle. t_1' does *not* equal $f(t_1)$; it is t_2' that corresponds, in this sense, to t_1.

reckoning of the other. In general $f(w)$ and $g(w)$ will not be the same function. It would obviously be more symmetrical if they were. And this is therefore what we set about achieving by altering the measures – recalibrating – either or both of the time scales of Red and Blue. We seek as it were a "functional square root", a function $h(\ )$ such that for any variable w, $h(h(w)) = g(f(w))$. We cannot do this directly, but we can find functions $h(\ )$ and $\psi(\ )$, such that for any variable x,

$$\psi^{-1}[h(h(\psi(x)))] = g(f(x)).$$

Red and Blue agree to recalibrate their clocks, so that henceforward, according to their new scheme of time reckoning, it will be true that a message dispatched by Red at t by Red's new time reckoning will be received by Blue at t' by Blue's new time reckoning, where $t' = h(t)$; and a message dispatched at t' by Blue's new time reckoning will be received by Red at t by Red's new time reckoning, where $t = h(t')$: and $h(\ )$ is the same in either colour.

By thus recalibrating their clocks, Red and Blue have gone as far as possible in achieving perfect symmetry between the one and the other, and eschewing egocentricity on the part of either. They have revised their systems of time reckoning so that now, even on the dates supplied by Blue, Red will not be able to say My messages travel faster than yours do, Blue – Мои письма идут быстрее ваших, Синий. Nor Blue to say My messages travel faster than yours do, Red – αἱ ἐμαὶ ἀγγελίαι θᾶττον τῶν σῶν, ὦ Πυρρέ, πορεύονται.

We now have developed two rules for ascribing dates at a distance. The radar rule, which applies to objects as well as persons; and what we may now call the *radio rule*, which applies only to other observers, who can communicate their own time reckoning in answering messages, as well as merely reflecting them. We thus have, essentially, two rules:

I (i) $t_2 = \frac{1}{2}(t_1 + t_3)$ the radar rule, depending on an arithmetic mean,
   (ii) $l_2 = \frac{1}{2}c(t_3 - t_1)$ giving the date and distance, assigned by Red, to a distant event, at Blue's home.
II (i) $t_2' = h(t_1)$ the radio rule, depending on a functional square root,
   (ii) $t_3 = h(t_2')$ correlating the date of dispatch, according to the time reckoning of the dispatcher, with the date of receipt, according to the time reckoning of the receiver.

Our problem is how Red is to correlate t_2 and t_2'; that is, to find a function $\phi(\ )$, such that $t_2 = \phi(t_2')$. I shall show that $\phi(\ )$ is a special case of L(iv) for when x' is $0'$ – i.e. when Blue says that he is at home $(0', 0', 0')$ – когда Синий говорит, что он дома We work out what $\phi(\ )$ must be if t_2 is to be $\phi(t_2')$ on all occasions – i.e. for all dates t and t' that are to be correlated by Red's transformation rule for translating red into blue. What we do, essentially, is to use the radar rule to express t_2 in terms of t_1 and t_3, and the radio rule to express t_1

and t_3 in terms of t_2'. We then can work out what $\phi(\ )$ must be if the radar and the radio rules are both to hold good for any and every t_2'.

By radar rule   I   (i) $t_2 = \tfrac{1}{2}(t_1 + t_3)$
By radio rule   II   (i) $t_1 = h^{-1}(t_2')$
By radio rule   II   (ii) $t_3 = h(t_2')$
$$\therefore \ \phi(t_2') = t_2 = \tfrac{1}{2}[h^{-1}(t_2') + h(t_2')]. \tag{1}$$

Since we do not know in general what $h(\ )$ might be, this is not enough for us to be able to say what $\phi(\ )$ is. We therefore need to use the second, main, radar rule, which correlates t_1 and t_3 with l_2, the distance. This is where we use the fact that Red is allowed to assume that Blue is moving with *uniform* velocity with respect to him. Together with the convention that Red and Blue have each agreed to set their zero dates at the time of their coincidence, it enables us to express l_2 again in terms of t_2, namely $l_2 = vt_2$. We should note that it is *not* necessary that Red and Blue should have coincided in time past, and actually have been in uniform motion ever since. All that is necessary is that they should be in uniform motion while they recalibrate their clocks and that each should set his zero to the date when, according to his calculation of the other's velocity, the other would have been, or (if the motion is towards each other) would be, coincident. This is enough. And of course we could derive our result, at the cost of somewhat heavy algebra, without making this simplifying assumption about the zero.

With this assumption we have, using radar rule I (ii),

$$t_2 = \frac{l_2}{v} = \tfrac{1}{2}\frac{c}{v}(t_3 - t_1)$$

and hence, as before,

$$\phi(t_2') = \tfrac{1}{2}\frac{c}{v}[h(t_2') - h^{-1}(t_2')]$$

$$\therefore \ \frac{v}{c}\phi(t_2') = \tfrac{1}{2}[h(t_2') - h^{-1}(t_2')]. \tag{2}$$

We can now solve for $h(\ )$ and $\phi(\ )$. It is convenient to solve for $h(\ )$ first. Adding (1) and (2)

$$\phi(t_2') + \frac{v}{c}\phi(t_2') = h(t_2')$$

$$\left(1 + \frac{v}{c}\right)\phi(t_2') = h(t_2'). \tag{3}$$

Subtracting (2) from (1)

$$\phi(t_2') - \frac{v}{c}\phi(t_2') = h^{-1}(t_2')$$

$$\left(1 - \frac{v}{c}\right)\phi(t_2') = h^{-1}(t_2'). \tag{4}$$

ERRATA

page 217, lines 16–17, the words and arguing in justification should be in black, not red.

page 222, equation (2) and following
page 223, equation (5)
page 223, equation 7 lines from end of page

In each case, the letter c should be in black, not red.

page 233, last equation. The first $\times$ should be in blue and the second $=$ sign in red instead of vice-versa.

Dividing (3) by (4)

$$\frac{1 + v/c}{1 - v/c} = \frac{h(t_2')}{h^{-1}(t_2')}. \tag{5}$$

This has to hold for any and every value of t_2'. It would therefore have to hold for w, where $w = h(t_2)$. Substituting w for t_2', we have

$$\frac{1 + v/c}{1 - v/c} = \frac{h(t_2')}{h^{-1}(t_2')} = \frac{h(h(t_2'))}{h^{-1}(h(t_2'))} = \frac{h(h(t_2))}{t_2'} \tag{6}$$

$$\therefore \ h(h(t_2')) = \left(\frac{1 + v/c}{1 - v/c}\right) \times t_2'. \tag{7}$$

It is easy to see that

$$h(t_2') = \sqrt{\left(\frac{1 + v/c}{1 - v/c}\right)} \times t_2'$$

is a solution; and, indeed, it can be shown to be the only one [6]. From it we can obtain $h^{-1}(t_2')$ and so $\phi(t_2')$

$$h^{-1}(t_2') = \frac{1}{\sqrt{\left(\dfrac{1 + v/c}{1 - v/c}\right)}} \times t_2' = \sqrt{\left(\frac{1 - v/c}{1 + v/c}\right)} \times t_2'$$

$\therefore$ by (1)

$$\phi(t_2') = \frac{1}{2}\left[\sqrt{\left(\frac{1 - v/c}{1 + v/c}\right)} + \sqrt{\left(\frac{1 + v/c}{1 - v/c}\right)}\right] \times t_2'$$

$$= \frac{1}{2}\left[\frac{1 - v/c + 1 + v/c}{\sqrt{(1 - v^2/c^2)}}\right] \times t_2'$$

$$= \frac{1}{\sqrt{(1 - v^2/c^2)}} \times t_2'$$

or, as it is more conveniently written,

$$(1 - v^2/c^2)^{-\frac{1}{2}} \times t_2'.$$

Hence, $t_2 = (1 - v^2/c^2)^{-\frac{1}{2}} \times t'$ is the way in which Red must correlate the date t_2 he ascribes by the radar rule to Blue's receipt of his message, with the date t_2' which he agrees, in virtue of the radio rule, Blue should assign to the same

[6]   If we differentiate (7) we have $h'(h(t)) \times h'(t) = a$, writing a for $(1 + v/c)/(1 - v/c)$ and, again substituting $h(t)$ for t, we have also $h'(h(h(t))) \times h'(h(t)) = a$. Hence $h'(t) = h'(h(h(t))) = h'(at)$, and, per contra, $h'(t) = h'(t/a)$ and similarly $h'(t/a) = h'(t/a^2)$. But if $v \neq 0$, $a > 1$, $1/a < 1$. Hence, for any t, the series $t, t/a, t/a^2, \dots$ etc. converges to 0. Granted that $h'(t)$ exists and is unique in the limit, it follows that for any t, $h'(t) = h'(0)$. $\therefore$ $h'(t)$ is a constant, and $h(t)$ must be of the form $At + B$, where A and B are constants, and so (8) is the only solution. I am indebted to G. J. Whitrow for pointing out the need for this lemma and providing an outline of the proof.

event of Blue's receipt of Red's message. Red must correlate t_2 and t_2' in this way if he is to have both the radar and the radio rule without inconsistency.

$\phi(t_2')$ is a special case of

$$L\text{(iv)} \quad t = \left(1 - \frac{v^2}{c^2}\right)^{-\frac{1}{2}}\left(t' - \frac{vx'}{c^2}\right).$$

We have $x' = 0'$, because Blue sends all his messages from Home, Οἴκοθεν, $(0', 0', 0')$. We have therefore worked out the transformation for the origin of Blue's frame of reference (or coordinate system), not for a general point (x', y', z', t'); but it is easy to generalize.

The key to the argument is that each person is *both* an *observer and* an *object*. He both is the centre of his own world, and occupies some position in the other's world. We have not one but two frames of reference, and every point of the one, including the origin, is a point of the other, and vice versa. Red can not only refer to Blue as an object, but tell Blue what he, Red, is saying about him, Blue, and be told by Blue, as an observer, how he, Blue, would describe the same events. Red then has to learn how to translate Blue's own account of himself, Blue, as an observer, into his, Red's, account of Blue as an object; and vice versa. The radar rule enables Red to talk about Blue: the radio rule enables him to talk with Blue about the same things. And the Lorentz transformations are generated from our making these two rules compatible.

§ 45
A priori arguments and empirical truths

The reader will be suspicious of the transcendental derivation of the Lorentz transformations in the previous section, and he will be wise. A fairly substantial rabbit – a contingent physical law – seems to have been plucked out of a very vacuous hat. And it is natural to suspect some legerdemain. Indeed, we may argue that there must have been some empirical assumptions smuggled in, or we should never be able to deduce any conclusions with empirical content. Such an argument is partly countered, but only partly countered, by the considerations of § 40. We need to press the question of what our recipes for easy symmetrical communications have to do with the detailed and exact experiments on which the Special Theory of Relativity is founded. The answer is that they provide a schema which might turn out to be of no use in our dealings with the physical world, but which, if they be instantiated, need no further explanation why they should be of this form rather than some other. It could be that when we had regraduated our clocks in order to secure the compatibility of the radio and the radar rules, we found they were no use for anything else. Happily, this is not so; but we concede that although we always *can* secure the Lorentz transformations by regraduating, it would be a pointless exercise, unless the resulting system for referring to dates and positions, and for measuring intervals and distances, were a useful one. The empirical content of the Special Theory of Relativity as here expounded lies in the fact that we can apply it to physical systems, and find that the time told by our regraduated system is the same as the time told by physical systems. It is an empirical fact, which could well have been otherwise, that the half-life of fundamental particles coming through the earth's atmosphere at very high speeds is, according to our earthbound reckoning, much greater than the half-life of similar particles at rest with respect to the earth. In spite of all our transcendental derivations of the Lorentz transformations, it remains an empirical question, to be decided by experiment, whether the laws of physics are invariant under the transformations; or, to put it the other way, whether those features that are unaltered by any Laurentian redescription are of any physical significance. The fact that they are is not a logically necessary one, in the sense of being a mere tautology. It is a fact, but a fortunate one; not merely contingent, but one that speaks well of the rationality of the physical world in which we dwell.

This fortunate fact is where the empirical content comes in. The Lorentz transformations are not only derivable *a priori*, but applicable to actual physical phenomena. They therefore do provide a schema which might, indeed, have turned out to be no use in our dealings with the physical world, but can in fact be applied. No further explanation is needed why they are of this form rather than some other. But for all their *a priori* attractiveness, it could have been that when we had regraduated our clocks to secure the compatibility of the radio and radar rules, we found they were no use for anything else. This is the risk, run by any account that starts from an abstract principle: the conclusions may follow from the premises, but may be irrelevant or inapplicable to the real world. So too if one starts, as Einstein did from the three premises that the laws of nature had the same form in all inertial frames of reference, that the velocity of light is independent of the motion of the light source, and that space is homogeneous [1]. Or, as he did later, from the first of these premises together with the invariance of ds^2 in all inertial frames of reference [2]. Or, if we argue from the first two of those premises together with the invariance of the direction of the ray of light which propagates parallel to the direction of the relative motion of the space coordinate systems [3]. Or (up to an arbitrary constant), if we argue from the invariance of

$$\frac{\partial^2 \phi}{\partial x_1^2} + \frac{\partial^2 \phi}{\partial x_2^2} + \ldots + \frac{\partial^2 \phi}{\partial x_n^2} = 0 \quad \text{(for } n \neq 2\text{) [4]}.$$

[1]   A. Einstein, *Annalen der Physik*, Bd., XVIII (1905), 898; tr. J. H. Barth in C. W. Kilminster (ed.), *Special Theory of Relativity* (Oxford, 1970), pp. 191 ff.
[2]   A. Einstein, *Vier Vorlesunghug über Relativitätstheorie*, XXI (1922).
[3]   Y. Memeesa and T. Iwatsuki, "On Linearity of the Lorentz Transformation", *Journal of Science, Hiroshima University*, A, I (1931), pp. 111–16.
[4]   V. V. Narliker, "The Restriction to Linearity of the Lorentz Transformation", *Proceedings of the Cambridge Philosophical Society*, XXVIII (1932), pp. 460–2.

Or in many other ways [5]. We can derive the Lorentz transformations from many different sets of assumptions, each one of which has considerable rational appeal, but still could be denied without self-contradiction. And although it is a further argument in favour of the Lorentz transformations that they can be derived from a number of different sets of highly plausible assumptions, there still remains the logical possibility, which the empiricist will make much of, that they will prove useless for talking about empirical phenomena. Both approaches are valuable, neither completely adequate; and it is one of the tasks of philosophy to show how *a priori* and empirical considerations are intermingled in constituting our concepts, so that time is neither, as empiricists assume, merely what the clocks say, nor, as philosophers are popularly supposed to believe, a matter to which all considerations of time-telling or punctuality are totally irrelevant. Both empirical and rational elements enter into our understanding of time and space. And if, as here, both approaches reach the same result, the proper conclusion is not that there has been some underhand cheating, but that we are in fact fortunate in living in, if not the best, at least the most rational, of all possible worlds.

[5]   For a discussion of the many different derivations of the Lorentz transformations, see G. Stephenson and C. W. Kilminster, *Special Theory of Relativity* (London, 1958), ch. I, esp. §§ 5–8, pp. 9–22.

§ 46
The dilatation of time

Many people have confused themselves about the implications of the Lorentz transformations. They are told that if they go very fast, spatial distances get shorter (the Fitzgerald contraction) and temporal intervals get longer (the dilatation of time). The former is difficult enough; the latter quite incomprehensible. The confusion, however, lies not in the phenomenon, but in our description of it. Length is not shortened nor time lengthened; but the rules for constructing a public time for different people at different – distant – places turn out to be more complicated than we had thought, if certain symmetries are to be achieved. Public time, as we have seen (§ 2), is a logical construct from private time. It is always possible that public time, and the clocks that measure it, seems to be running fast or slow by comparison with the private time of our own experience. What the Lorentz transformations show is that this must be the case if we seek to establish a public time between people at great distances from each other, moving at great (though uniform) velocities with respect to each other. Since they are at different places, they do not have a simple uncomplicated intersubjective experience, on which public time, and a system for measuring time, can easily be based. To establish any intersubjective experience requires some artifice, and to establish a metric that is adequately non-arbitrary is even more difficult. Especially if people are not only far apart, but moving, their time scales are naturally, or at least easily, discordant. The Lorentz transformations are the best way of bringing them into step, and the Fitzgerald contraction and the dilatation of time are the cost at which we purchase our re-established harmony between different centres of consciousness, differently situated and differently moving, each with its own viewpoint, but each communicating with others and enjoying parity of epistemological esteem.

Let us consider $L(\text{iv})$, in the special case for when $x' = 0'$, namely

$$t = \left(1 - \frac{v^2}{c^2}\right)^{-\frac{1}{2}} t'.$$

As we have seen, if $v^2 > 0$ (i.e. if $v \neq 0$)

$$\left(1 - \frac{v^2}{c^2}\right)^{-\frac{1}{2}} > 1.$$

It follows that unless $v = 0$, $t > t'$. This is often expressed, in a rather confusing way, by the statement that Blue's clock is running slow, or, worse, that time slows down as v increases. What we want to say is not that Blue's clock is running slow but that Red might be tempted to say Your clock is running slow, Blue – Ваши часы отстают, Синий. The temptation to say this is simply that $t > t'$, so that Blue's time reckoning will register earlier dates and the elapse of smaller intervals than Red's radar time reckoning will. So that if Red were utterly and egocentrically confident of his own radar time reckoning, he would blame the discrepancy on Blue's system of time reckoning. To do this would be to contravene the canon of non-egocentricity. It would also be to abandon the radio rule. The communication argument satisfies two purposes together: it fulfils a purpose of non-egocentricity, enabling me to allow for, and give due respect to, the temporal experience of other minds, even distant observers, who can observe my signals and reciprocate them; it also satisfies a criterion of symmetry, that the signal function $h(t)$ should be the same both ways, i.e. that $t_2' = h(t_1)$ and $t_3 = h(t_2')$. Red ascribed t_2 to the distant event of Blue's receiving his message; but in an important sense the date is also partly assigned by Red, in that he and Blue agree to recalibrate their clocks in order to ensure that the signal function shall be the same both ways. Without consulting Blue, Red might have set about determining a signal function $h(t)$, such that $t_3 = h(h(t_1))$, and ascribed to Blue the 'local' time $t_2' = h(t_1) = h^{-1}(t_3)$. This might have shown unwarranted confidence on the part of Red about the reliability of his own clocks. But it does show also how the Lorentz transformation is generated by our having two symmetry conditions: one in the radar rule, that $\epsilon = \frac{1}{2}$, i.e. that t_2 shall be the arithmetic mean of t_1 and t_3, the other in the radio rule that the signal function shall be the 'functional square root' of the function correlating t_1 and t_3, i.e. that $t_2' = h(t_1) = h^{-1}(t_3)$. Given these two symmetry conditions, we are likely to get discrepancies between t_2 and t_2'; but we should not let Red blame them on Blue's clock running slow – τὴν τοῦ κυανέου κλειψύδραν βραδέως ῥέουσαν but on Red himself using both his radar rule and his radio rule. The Lorentz transformations are the price of serving two masters of symmetry.

The dilatation of time is not some incomprehensible slowing down of time, but an obscure way of saying that, for very good reasons, we have two ways of assigning dates to distant events, and of measuring intervals in distant places, and that the one which we feel ought to be considered as the "local time" of a distant event assigns earlier dates and shorter intervals than the one which we regard as being peculiarly "our own" time. This is teasing, but it is not much more teasing than the fact that on the earth's surface we assign "local time" to distant places which is different from "our own" time. Noon by San Francisco time is late evening by G.M.T., and occasionally this can raise difficulties [1]. The Lorentz transformation is one degree more difficult, because there is not only a change in the *dating* system, but also a difference in the *metric*, the system of measurement. We need to remember that for distant events and distant places any scheme of measurement is an assigned one. There is no common naturally given measure of time shared by Red and Blue. Each must measure temporal intervals as best he can, and then by comparing notes, at long range, construct a system of measurement that is as intersubjective as possible. If we find differences between different ways of assigning dates and intervals, this shows something about our conceptual system, and may or may not be a criticism of it; but it shows nothing to the effect that time itself is being slowed down.

One particular difficulty deserves further notice. People are often puzzled how the dilatation of time can take place both ways. If Blue's clocks seem to be running slow by Red's reckoning, surely then Red's clocks must be running fast by Blue's reckoning, not slow as the relativity merchants claim.

The resolution of this problem turns on the twofold distinction between the direction of translation and the person by whom the translation is being done. If we were simply altering the first of these, we should have to replace *slow* by *fast*. If Blue's clocks seem to be running slow, i.e. slower than Red's clocks, by Red's reckoning, then Red's clocks seem to be running fast, i.e. faster than Blue's clocks, by Red's reckoning. But this does not hold if we interchange *both* the direction of translation *and* the person by whom the translating is being done.

[1]   See R. *v*. Logan and Others, Courts Martial Appeal Court, 21 June 1957, *All England Law Reports* (1957), II, pp. 688–90. The Army Act, a statute enacted by the Queen in Parliament, was valid for Hong Kong as well as the U.K. The statute was to come into force on 1 January 1957. Was the statute in force in Hong Kong in the early hours of 1 January 1957, by Hong Kong time, when at Westminster it was still 31 December 1956 and the statute was still inoperative?

We need, therefore, always to be asking *where* the comparison is being made. If Red is comparing Blue's dates with the dates that he, Red, ascribes to Blue by the radar rule, he will say Your time reckoning, Blue, ascribes earlier dates and shorter intervals than mine does, i.e. $t' < t$, and I feel tempted to say that your clocks are running slow, Blue – Ваш расчёт времени приписывает более ранние даты и более короткие промежутки времени, чем мой расчёт, $t' > t$, так что я склонен сказать, что все ваши часы отстают. If Blue is comparing Red's dates with the dates that he, Blue, ascribes to Red by the radar rule, he will say Your time reckoning, Red, ascribes earlier dates and shorter intervals than mine does, i.e. $t < t'$, and I feel tempted to say that your clocks are running slow, Red – Κατὰ τὸν σὸν τῶν χρόνων λογισμόν, ὦ Πυρρέ, πρωιαίτεραι μὲν ἀποβαίνουσιν αἱ ὧραι, βραχίονα δὲ τὰ διαστήματα, τοῦτ᾽ ἔστιν $t < t'$· ὅθεν κινδυνεύω λέγειν ὅτι βραδέως ῥέουσιν αἱ σαὶ κλεψύδραι, ὦ Πυρρέ. But these two statements do not contradict each other because they are made about ways of dating *different* events, different because at different places. Red is talking about dates at Blue's home, Blue is talking about dates at Red's home. If Red could get Blue to talk about the dating of *the same events* as Red was talking about, Blue would agree with Red, namely Your time reckoning, Red, ascribes later dates and longer intervals to events happening here, at home $(0', 0', 0')$, than mine does, i.e. $t > t'$, and I should feel tempted to say that your clocks were running fast, Red, if I were tempted to say that sort of thing at all. But, of course, I do not think it is anything to do with your clocks, but only a feature of your way of ascribing dates to events that are happening here – ὁ σὸς τῶν χρόνων λογισμός, ὦ Πυρρέ, τοῖς ἔνθαδ᾽ οἴκοι $(0', 0', 0')$ γιγνομένοις ὑστέρας μὲν τὰς ὧρας προσάπτει, μακρότερα δὲ τὰ διαστήματα, ἢ ὁ ἐμός, τοῦτ᾽ ἔστιν $t > t'$· ὅθεν κινδυνεύοιμ᾽ ἂν εἰπεῖν ὅτι αἱ σαὶ κλεψύδραι, ὦ Πυρρέ, ταχέως ῥέουσιν, εἴ τί που καὶ κινδυνεύοιμι τοιαῦτα λέγειν. ἀλλ᾽ οὐ γὰρ οἶμαι τοῦτο ταῖς σαῖς κλεψύδραις οὐδὲν προσεῖναι, ἀλλὰ σοὶ μόνῳ, ὁποῖον τρόπον τῶν ἔνθαδε $(0', 0', 0')$ γιγνομένων τοὺς χρόνους ὁρίζεις. The reason why this point is not obvious is that our rules for ascribing dates are all rules for ascribing dates to *distant* events. We do not feel the need for ascribing dates that happen at home, because we feel we know when they happen, because we are there when they occur. Thus Blue does not think about the dates that Red is ascribing to him – why should he? He is quite happy with his own dating system. The air of paradox arises because each person is busy minding the *other*'s temporal business without realizing that this is what he is doing. The idea of a clock, which Einstein used to define *local* time, but which we think of as measuring all time, adds to the confusion. Red feels that grass elsewhere is greener, the pace of life more leisured, and even the clocks stand almost still. But Blue feels just the same, and has an equal urge to epistemological wanderlust.

We tend to use dictionaries only when abroad. We find it unnatural to think in a foreign language when we are at home, and difficult to make all the adjustments that are required. It is, however, possible, if we take care to make the necessary adjustments. So with the Lorentz transformations. The air of paradox arises from the conjunction of the two rules

$$L \text{ (iv)} \quad t = \left(1 - \frac{v^2}{c^2}\right)^{-\frac{1}{2}} \left(t' + \frac{vx'}{c^2}\right)$$

$$\Lambda \text{ (iv)} \quad t' = \left(1 - \frac{v'^2}{c^2}\right)^{-\frac{1}{2}} \left(t + \frac{v'x}{c^2}\right)$$

or, equivalently, the two rules

$$N \text{ (iv)} \quad t' = \left(1 - \frac{v^2}{c^2}\right)^{-\frac{1}{2}} \left(t - \frac{vx}{c^2}\right)$$

$$M \text{ (iv)} \quad t = \left(1 - \frac{v^2}{c^2}\right)^{-\frac{1}{2}} \left(t' - \frac{v'x'}{c^2}\right).$$

Since $(1 - v^2/c^2)^{-\frac{1}{2}} > 1$ and $(1 - v'^2/c^2)^{-\frac{1}{2}} > 1$ for $v \neq 0$, $v' \neq 0$, it seems that in every case there is a factor, β, greater than unity (see p. 210), so every time we translate t' into t (by N (iv) or by Λ(iv)) or t into t' (by L(iv) or by M(iv)) we multiply by β, so that if we were to translate and retranslate back again the final result would be β^2 times as great as the original. But in reaching this paradoxical conclusion we have forgotten to consider the values of x and x', which tell us where the event is whose date we are discussing. We have unconsciously assumed that the values are always 0 and 0'. This is true when Red is using L(iv) to translate red dates into blue dates for events at Blue's home; and similarly if Blue is using Λ(iv) to translate blue dates into red dates for events at Red's home. But it is not true if Red were to use L(iv) to translate dates of events at his own home into blue dates. For then x' does not have the value 0', but, thanks to the way Blue has fixed his zero date, $v't'$. Similarly if Blue were to use Λ(iv) to translate dates of events at his own home into red dates, he must not assume $x = 0$, but $x = vt$. Putting in these values for the position will just outweigh the effect of the factor β.

Thus L(iv) reads

$$t = \beta\left(t' + \frac{vx'}{c^2}\right).$$

If $x' = v't'$,

$$t = \beta\left(t' + \frac{vv't'}{c^2}\right) = \beta\left(1 + \frac{vv'}{c^2}\right)t'.$$

Since $v' = -v$

$$t = \left(1 - \frac{v^2}{c^2}\right)^{-\frac{1}{2}}\left(1 - \frac{v^2}{c^2}\right)t' = \left(1 - \frac{v^2}{c^2}\right)^{+\frac{1}{2}} \times t'$$

which is just what we need, if the rule Λ(iv) is also to apply. That is, if exceptionally, Red uses L(iv) for events at Red's home, Дома, $(0,0,0)$ he must remember that his home is, to Blue, the position $(v't',0,0)$, and he will therefore obtain by feeding in these position coordinates into L(iv) the rule, for this special (and exceptional) case

$$t = \left(1 - \frac{v^2}{c^2}\right)^{+\frac{1}{2}} \times t'$$

which exactly fits Blue's rule Λ(iv), applied more normally by Blue to distant events, i.e. events at Red's home. For Λ(iv) reads

$$t' = \left(1 - \frac{v'^2}{c^2}\right)^{-\frac{1}{2}} \left(t + \frac{v'x}{c^2}\right)$$

which becomes, when $x = 0$,

$$t' = \left(1 - \frac{v'^2}{c^2}\right)^{-\frac{1}{2}} \times t$$

which could also be expressed

$$t = \left(1 - \frac{v'^2}{c^2}\right)^{+\frac{1}{2}} \times t' = \left(1 - \frac{v^2}{c^2}\right)^{+\frac{1}{2}} \times t'.$$

Thus the Lorentz transformations do not yield any inconsistent or paradoxical results. The factor $\beta = (1 - v^2/c^2)^{-\frac{1}{2}}$, which is always greater than unity unless $v = 0$, does indeed make "local times", determined by the radio rule, appear to be "slow" by comparison with the "observer's own time" assigned by the radar rule, because in applying the Lorentz transformations to "local time" we have also to have "local places", i.e. a position referred to in a frame of reference whose origin is that position, i.e. a position whose spatial coordinates will be $(0,0,0)$. If Red wants to talk of Blue's home time, he means that he wants to talk of the dates t', which Blue assigns to events at Blue's home $(0',0',0')$. And since he wants to have $x' = 0'$, he will find that his, Red's, assigned dates t to events at Blue's home will be greater than Blue's dates t' by a factor β. But "distant times" are correspondingly "fast", because in applying the Lorentz transformations to distant places, the spatial coordinates will *not* be $(0,0,0)x$; the effect of the spatial component x in the Lorentz transformation will outweigh the effect of the factor β. If Red wants to talk of his own home time as being Blue's distant time, he must remember that x' will be quite large, and will make the dates t' that Blue assigns to τοῖς πόρρωθεν γιγνομένοις, distant events at Red's home, greater than his own dates t, so that Blue's distant dates could be said to be "fast" with respect to Red's own dates.

It might be thought that the Lorentz transformation was a heavy price to pay for having both the radio and the radar rules. Rather than have the complication of different rules and different ways of reckoning dates and measuring intervals, should not we do better to have only one rule, and stick by that?

But we do need both rules. We need the radar rule if we are to establish a spatio-temporal framework – if we are to ascribe times to distant *places* (only, of course, we cannot locate distant places directly, and we have to think of them as occupied by distant objects). We need the radio rule if we are to allow for the independence of *other things* (only, of course, in order to discuss them, they must have temporal properties independently of us, and so we have to think of them as other persons, observers). If we had only the radar rule, there would not be parity of esteem between all observers. If we had only the radio rule, we would have an idea of other-ness, but none of distance, and in fact would be, unwittingly, ascribing different dates to the receipt of the same radio message by two observers in the same place, but moving at different velocities. There would be no idea that change would ever come to anything, that two observers could ever meet.

The Special Theory of Relativity is relativistic only with regard to dates, places, orientations and velocities, not accelerations or rotations. If two observers have a non-uniform motion with respect to each other, there is no reason why they should enjoy parity of esteem. One or the other or both has undergone a force, which may well have affected the running of his clock. Therefore, in a flat, Euclidean space, we do not have to worry about the "clock paradox", in which two observers synchronize their clocks, separate at a great speed, come round and meet again, and wonder whose clock has gone wrong. For at least one, and perhaps both, must have been accelerated, and undergone a force. If you go pot-holing, and I go sailing in a dinghy, and we find our watches not telling the same time afterwards, we are not greatly surprised. Nor is there a "real time" that our watches, if better, would have told correctly. All measurements of time are to some extent conventional, to some extent arbitrary. We can lay down certain standards of non-arbitrariness, that the measure of time shall be invariant under certain transformations. But we cannot have it invariant under all transformations. How could we? And if we have it invariant under some, it will not be invariant under others. In particular, if we seek a Laurentian harmony we shall have discordant measures between accelerated observers, just as we should between observers in instantaneous (perhaps telepathic) communication with each other (see § 42, p. 198).

The clock paradox is thus a seeming paradox only, so far as the Special Theory of Relativity goes. Natural laws are not invariant in accelerated frames of reference, nor are clocks unaffected by forces. If an astronaut set out and went for a long voyage through the universe at great speed, and then returned again, 100 years later by our reckoning, but claiming that only a year had elapsed by his, we should regard his case in much the same way as if his bodily processes (together with his watch) had been slowed by being cooled down. It would be possible for a man to meet his remote descendants – but at the cost of having been put in cold storage for the intervening period, and therefore not having done much. Similarly the astronaut would have been "out of circulation" for a century. He would be able to meet his great-great-grand-children, but would not have done much in the intervening years. He would have achieved only one year's worth of work, and, if an academic, would find many of his conclusions anticipated and sadly out of date. By earth standards his mind would have moved very slowly over the last hundred years. He would have been almost as much in a state of suspended animation as if it had been induced by hypothermia. He would be a Rip Van Winkle, a survival from an earlier age. Of great interest to the historian, but not a problem for a philosopher.

§ 47
The Special and General Theories of Relativity

Much confusion is engendered by the Special and General Theories of Relativity. It is partly the word 'relativity'. "Relative to what?" we may ask, and be a long time before getting an intelligible answer. It is partly also that their names suggest that the Special and General Theories of Relativity are fairly similar in approach and conclusions, and that both are to be contrasted with Newtonian notions of absolute space and time. But in many respects the affinities are between the Special Theory and Newtonian mechanics, while the General Theory is the one that is radically different in inspiration and results.

All scientific theories are relative. They explain some features and ignore others. And in laying down canons of relevance, they implicitly specify relative to what respects they require situations or systems to be described. In particular, the principle of date-indifference stipulates that it shall not matter how we date the various stages of a physical process, provided we do not alter the magnitudes assigned to temporal intervals. We can transform from one dating system to another with a different zero (say from A.U.C. to A.D.), and the description of natural laws and physical processes will be unaltered (see § 12, pp. 71–2). In a similar fashion the principle of origin-indifference and orientation-indifference requires that it shall not matter so far as physics is concerned if we redescribe a system in a frame of reference with a different origin or with axes in different directions. The laws of physics are invariant under a transformation that merely alters the temporal or spatial origin, or reorientates the spatial axes. Physics is concerned only with the equivalence class of all the acceptable redescriptions of a situation or a system; and if we choose, as we do, one description rather than another, we need to be very clear that the description we are using is relative to a frame of reference which is arbitrary inasmuch as there is no physical reason for choosing it rather than any other. So far as physics is concerned, the choice between A.U.C. and A.D. is entirely arbitrary, although we may regard our reasons for adopting one of these dating systems as adequate.

All scientific theories are absolute. They recognize some features as being of importance, and coin descriptions of them which do not vary from one standpoint to another. In applying these descriptions there is no need to specify which system is being used. The chemist does not have to measure atomic weight in grammes or ounces, and two lines are perpendicular whatever Euclidean axes are used. It follows that not every redescription is acceptable. It is not acceptable to regraduate time so that what were isochronous intervals no longer are (§§ 10–14), nor to operate with a non-Euclidean geometry when we are dealing with simple kinematics. In brief, transformations that leave unaltered the description of important features or relations or laws are acceptable, and transformations that do not leave them invariant are not acceptable. If we know what is to be invariant, we can tell what range of redescriptions are acceptable, just as earlier we could argue from the admissible redescriptions to the significant features (§§ 30, 31).

Every scientific theory is both relative – because not every feature is relevant to it – and absolute – because some features are. Newtonian mechanics and the Special Theory of Relativity are relative in that they are

    (1) origin-indifferent with regard to time (date-indifferent),

    (2) origin-indifferent with regard to space (position-indifferent),

    (3) orientation-indifferent,

    (4) unaltered by spatial reflection,

    (5) unaltered by reversal of time,

    (6) unaltered by a constant velocity.

And they are absolute in that they are affected by

    (a) a change of scale (magnification),

    (b) an acceleration,

    (c) a rotation (angular velocity).

We have given *a priori* arguments in favour of (1) in §§ 10, 12 and 14 and of (2) and (3) in § 35, and have accepted (4) in § 36. (5) is something of a blemish, and points to some inadequacy in Newtonian mechanics and the Special Theory of

Relativity, which we shall discuss later. It is well known that (6) is characteristic of the Special Theory of Relativity, but we need to reiterate that it is also characteristic of Newtonian mechanics: if we change from one frame of reference to another moving at a constant velocity with respect to it, all the laws of Newtonian mechanics will come out unaltered.

Of the transformations that do not leave everything unaltered, (*a*) needs only a little comment. The frequency and wavelength of, say, sodium light are physically determinate. It would make a difference if everything were twice as large, or things happened twice as fast. Within classical Newtonian mechanics, the constant of gravitation would be altered by any change of temporal or spatial scale. This is not to say that our *units* of measurement are naturally given. Some are (angles, the velocity of light, Planck's constant); most are not. But physics in general, unlike Euclidean geometry (see § 38), is not scale-invariant.

The other two transformations (*b*) and (*c*) bear the same resemblance to each other as (3) and (6) do (see §§ 42–3). Newton took (*c*) as the typical transformation that did not leave the laws of physics unaltered, and argued that we could tell whether a bucket full of water possessed an absolute angular velocity or was absolutely at rest, so far as rotation was concerned, by seeing whether its surface was concave or flat [1]. Clarke based his argument on (*b*) and maintained that an indiscernible motion of the universe would nevertheless, upon a sudden stop, have real effects [2].

Newton's and Clarke's arguments do not show, as they thought they did, that space is absolute; but only that one argument for its being relative is invalid. This argument is based on the interplay of dates and intervals, positions and distances, directions and angles, that we noted earlier (§ 11), and argues that, since we use real numbers as labels to refer to dates, positions and directions, we must necessarily refer to them by measuring the interval, distance or angle between them and some arbitrarily adopted origin. Newton and Clarke show that this argument cannot be valid, for angular velocities and accelerations are equally measured by real numbers, and are thus relative to some frame of reference, but are absolute so far as Newtonian mechanics is concerned. The fact that our standard system of reference refers to dates and positions by measuring intervals and distances from some arbitrary origin is inconclusive. In fact we can, as we saw, refer to dates, positions and directions in other, less systematic but still usable, ways (§ 11); and in any case there might be real physical effects that led us to prefer one frame of reference to another (see § 14, p. 81, and n. 4; also § 42, pp. 197–8).

Mach disputed this. He denied the force of Newton's bucket experiment, saying that the concave surface of the rotating water could be attributed to the

[1]   *Principia*, Scholium to Definition VIII, § IV; reprinted in H. G. Alexander (ed.), *The Leibniz-Clarke Correspondence* (Manchester, 1956), pp. 157–8.
[2]   Fourth Letter to Leibniz, § 13; in Alexander, op. cit., p. 48.

effect of the stars rotating round the bucket [3]. It could. Any number of *ad hoc* hypotheses could be adduced to explain a given phenomenon; and some are not susceptible of direct empirical falsification but are none the less implausible for that. If Mach's alternative account were put forward seriously, we should have plenty of reason for rejecting it – we do not think the stars could suddenly come to have velocities far greater than that of light. But in any case, Mach's argument misses the point of Newton's bucket experiment, which is not to establish conclusively the whole of Newtonian mechanics, but to show that there is some discernible difference between two systems which may, in the light of Newtonian mechanics as a whole, be plausibly attributed to one system having an absolute angular velocity and the other not. In Newtonian mechanics and the Special Theory of Relativity there should be real effects of accelerations and angular velocities, and we should be able to distinguish un-accelerated and non-rotating frames of reference from all others. So far as these two theories go, we cannot distinguish frames at rest from other frames in uniform motion, nor those with one orientation or temporal or spatial origin from those with others. But it still makes sense to ask whether some one frame may not be preferable to others, and, as we have seen, there could be reasons for distinguishing one temporal origin from all the others [4].

So far as Relativity is concerned, Newtonian mechanics and the Special Theory of Relativity are very similar. They are relative as regards temporal and spatial origins, orientation, reflection and time reversal, and not relative as regards magnification, acceleration and angular velocity. The differences between them are that Newtonian mechanics is operating with time and space as completely different from each other, whereas the Special Theory of relativity operates with a unified space-time; and in consequence Newtonian mechanics admits infinite velocities and action at a distance, whereas the Special Theory of Relativity has its fundamental velocity finite; with the further result that Newtonian mechanics applies only to forces and matter whereas the Special Theory of Relativity can accommodate wave phenomena and electromagnetism as well. The General Theory of Relativity is like the Special Theory in operating with a unified space-time, but it starts from very different assumptions and has very different aims. The General Theory seeks to obviate the need for the concept of force, particularly gravitational force, by making all physics a part of geometry. Instead of having a world described by a flat Euclidean

[3]   Ernest Mach, *The Science of Mechanics*, tr. T. J. McCormack (Chicago, 1902), ch. 2, § 6, pp. 229–38.
[4]   For a general account of the bucket experiment, see Herman Erlickson, "The Leibniz-Clarke Controversy", *American Journal of Physics*, XXXV (1967), pp. 89–98; or P. G. Bergmann, *Introduction to Theory of Relativity* (Englewood Cliffs, 1942), Introduction; C. D. Broad, *Scientific Thought* (London, 1927), pp. 108–9; E. Nagel, *The Structure of Science* (New York, 1961), pp. 208–11. See also R. P. Feynman, R. B. Leighton and M. Sands, *The Feynman Lectures on Physics* (Reading, Mass., 1966), vol. I, § 16.1, for an illuminating discussion and a very proper rebuke to "cocktail-party philosophers" who argue for relativity.

geometry *and* various fields of force, the General Theory of Relativity describes the world in terms of a geometry that has an intrinsic curvature, which itself accounts for the curved paths taken by free particles without the need to postulate any sort of force. In Newtonian mechanics the flight of a cricket ball, and in the Special Theory of Relativity the path of an electron in a magnetic field, are described as being curved because of the effect of certain forces: in the General Theory of Relativity the paths are described as *soi-disant* straight lines – "geodesics" – which appear curved not because of any force but because space-time is itself curved.

The General Theory of Relativity is thus both much more ambitious and much more limited than either Newtonian mechanics or the Special Theory of Relativity: more ambitious because it seeks to explain more; more limited because, in order to explain fields of force in terms of local curvature, it has to forgo the principles of origin- and orientation-indifference which are fundamental to Newtonian mechanics and the Special Theory of Relativity. All places and all times are alike on the standard scientific view of space and time, and any difference between one spatio-temporal position and another is to be attributed not to the fact that they are spatially or temporally different but to some *other* factor – a difference of magnetic field or a change of temperature. Space and time can be flat, homogeneous and altogether featureless just because any feature that characterizes one spatio-temporal position and not another will be attributed to some factor other than space or time. But the General Theory of Relativity cannot secure the featurelessness of space and time by these means because it aims itself to account for all dynamic phenomena without positing any further forces or fields. Since evidently there are differing forces – or at least differing accelerations of freely moving particles – in different places and at different times, it follows that the General Theory of Relativity cannot regard all dates and all positions as being alike, but must ascribe differing spatio-temporal properties to different spatio-temporal positions *as such*. The General Theory of Relativity cannot maintain the six principles of invariance that are characteristic of Newtonian mechanics and the Special Theory of Relativity. If we change the origin, it will make a difference to the form taken by natural laws. Instead of principles of *in*variance, the General Theory of relativity lays down a principle of *co*variance, which states, in effect, that whatever frame of reference we adopt, certain differential properties of laws of nature will be preserved: to be exact, their "covariant derivative" has to be independent of the particular labelling system employed [5].

The General Theory of Relativity allows a multitude of different descriptions of space-time, and enables us to pick out very subtle samenesses. But it lacks the crude symmetries of date-indifference and position-indifference we were led to demand of time and space in our efforts to achieve a rational theory of clocks and to have space analogous to time. We could still have clocks – indeed

[5]   See for example, P. G. Bergman, op. cit., p. 15.

we could still have a rational theory of clocks based on the deep covariances of the General Theory of Relativity instead of the simple principle of date-indifference. But it is a pre-Newtonian theory; indeed, as regards space, it is Aristotelian. Every point has its own properties, described by the world tensor, which are characteristically different from those of other points; and if we want to know why a thing moves as it does, the answer essentially is because it is where it is. Space is no longer the container in which things are moved by forces, time no longer the container in which events are produced by causes; rather, space-time itself determines the movement of things, itself is the cause of events. Space-time, instead of being the featureless background, in which things exist and which makes just enough difference for them to be numerically distinct but not enough for them to be *eo ipso* qualitatively different, now becomes the highly structured entity that explains all mechanical and electro-magnetic phenomena. It is no longer a generalized, and extremely thin, adjective, but a very substantial noun. Moreover, we are led to regard space-time as a single entity, rather than a manifold of separate positions, since the characterization of each position depends on the frame of reference, which refers not to it alone but only along with every other one. Space-time is not only a noun rather than an adjective, but the one and only noun referring to the one and only substance, the unique *deus sive natura*, which describes, accounts for and explains all phenomena. As against Newton, the Special Theory of relativity embodies many of the insights of Leibniz, and reveals a space-time inhabited by harmonious monads, instead of a space in which impenetrable atoms at times collide. But in the General Theory of Relativity we become monists, rather than pluralists, in the philosophical tradition of Spinoza rather than of Leibniz or of Newton and Locke. It is an austere, all-embracing system, which views the whole universe as a unified self-subsistent substance, a single process developing according to its own inherent principle.

§ 48
Athanasius intra mundum

We saw earlier that space – philosopher's space – must have at least two dimensions (§ 28, pp. 133–4; § 35, p. 166). In fact, we all know that it has three. But we find it very difficult to see why there should be just three, and equally difficult to leave it as a brute contingent fact, for which no explanation can be given nor should be sought for. We are fairly sure that it is a synthetic truth – mathematicians use spaces of any number of dimensions without inconsistency, and clever people play four-dimensional chess with apparent pleasure, and even we use a two- or a four- or a five-dimensional space for some special purposes. Yet although we can imagine – just – what a two-dimensional world would be like, we find a four-dimensional space altogether inconceivable, and feel that the fact that our space is a three-dimensional one is no *a posteriori* truth which happens to be so, but might well have been different [1]. It feels as though it were necessarily so, and that nothing could shake our trust in the tridimensionality of space. As regards God's *sensorium*, if not His *personae*, we are all orthodox trinitarians; but we lack a reasoned defence of our faith.

[1] But see H. Brotman, "Could Space be Four Dimensional?", *Mind*, LXI (1952), pp. 317 ff.

One reason [2] why space needs to be three-dimensional is because three is the next larger number than two, and two is excellent Pythagoreanly (see § 39, pp. 185–7) and on the score of reversibility (see § 49, p. 253). If V is the potential in an n-dimensional space, we express the geometrical properties assigned to it by Laplace's equation [3]

$$\frac{\partial^2 V}{\partial x_1^2} + \frac{\partial^2 V}{\partial x_2^2} + \ldots + \frac{\partial^2 V}{\partial x_n^2} = 0.$$

"Its algebraic form (sum of second-order derivatives)", to quote Whitrow, "expresses the Pythagorean distance law of Euclidean space, and the fact that there are three independent variables, x, y and z, corresponds to the number of spatial dimensions postulated. It can easily be verified that an equation of Laplace's type in n independent variables would correspond to an inverse $(n-1)$th power law of force in a quasi-Euclidean space of n dimensions." [4] The inverse square law is therefore the appropriate law for attraction and repulsion in a three-dimensional space, if Laplace's equation is to hold. We can see that the inverse square law is reasonable if we consider that the surface of a sphere of radius r is $4\pi r^2$, so that if we are to conserve energy, we need the density of field per unit area to be inversely proportional to r^2. Therefore if we value the inverse square law, we must have space three-dimensional. But why should we have the inverse square law? Could not we have an inverse cube law, or for that matter a simple inverse law with gravitational and electromagnetic attraction varying with $1/r$?

Whitrow argues against the inverse cube law, and all laws based on negative indices of greater absolute magnitude, on the grounds that with them orbits are unstable. Only if the gravitational attraction varies with r^{-n}, where $n = 1$ or 2, can we have a theory of Newtonian orbits in which planets circle round the sun year after year. But unless the earth had a stable orbit, so that the sun's heat did not vary very greatly from time to time, life would be impossible. If the earth did not go in any orbit at all, or if it went in very eccentric elliptical orbits, like the comets, ice ages so intense that the oxygen of the atmosphere would liquefy would be separated by torrid epochs in which the rocks would melt. It is therefore an *ecological* necessity that space should not have more than three dimensions, because else human life, and hence human thought, would be impossible. It would be different for angels or any other non-corporeal intelligences – perhaps that is why Henry More maintained that spirits have

[2] This argument and the following ones are due to G. J. Whitrow, "Why Physical Space has Three Dimensions", *British Journal for the Philosophy of Science*, VI (1955–6), which is strongly recommended to the reader. It was partly anticipated by Kant. See his "Thoughts on the True Estimation of Living Forces", tr. J. Handyside, Kant's *Inaugural Dissertation and Early Writings* (Chicago, 1929), §§ 9–10, pp. 10–12.
[3] This expresses the condition that the value of V at any point is equal to the average value for the neighbourhood. It is a somewhat egalitarian condition imposed on the points of space with regard to the potential they are allowed to have. Local egalitarianism, so to speak.
[4] Whitrow, op. cit. pp. 28–9, slightly altered.

four dimensions [5]. But with our more earthy natures, it is a precondition of our having any experience at all that we live in a two- or three-dimensional world.

But why three rather than two? Whitrow offers two arguments, which I shall call the low-don argument and the high-don argument. The low-don argument turns on the fact that continuous physiological processes are possible only to an organism that is, topologically speaking, a torus, and that no torus can exist in a purely two-dimensional world. A two-dimensional organism could respire – it could breathe in and then breathe out; or it could, like an amoeba, ingest food and then expel the remains. But it could not have an alimentary canal, because this, in two dimensions, would divide the amoeba in two. Only in three dimensions can we have an alimentary canal, only in three dimensions can life be one continuous guzzle.

This argument lacks conviction. High table is an ornament to, rather than an essential condition of, academic life. If amoebae can, why should not we. Life might be less convenient, less elegant, if based on batch processes rather than continuous ones; but it would still be possible, and whatever the news-papers say about the colleges of Oxford and Cambridge, it cannot be a neces-sary truth that one has to be a gourmand to be a thinker. Far more convincing is the high-don argument [6], which points out that a cerebral network in two dimensions would be necessarily extremely limited, whereas in three dimensions there is no topological limit to its complexity. The Königsberg bridge problem shows how few different paths would be available to a neural network in only two dimensions [7]. If we are to have brains complex enough to enable us to discuss the dimensionality of space, we must live in a world of at least three dimensions. But our mortal clay could not survive, certainly could not have evolved, in a world of more than three dimensions. Therefore exactly three.

Whitrow's arguments apply more to Newtonian space than to a plenum. Although he invokes Laplace's equation and the concept of a field, his argu-ments turn on the possibility of planetary orbits under non-Newtonian gravi-tational laws, and the fact that we have bodies, and our intelligences can only function on condition of our being a certain sort of thing. We now turn to a more Leibnizian argument based not on the conditions required for human

[5]   Henry More, *Enchiridion Metaphysicum* (London, 1671), pt. I, ch. XXVIII, § 7, p. 384.
[6]   Originally due to J. B. S. Haldane; given by G. J. Whitrow, *The Structure and Evolution of the Universe* (2nd ed. London, 1959), p. 200, and brought to my attention by Mr A. Slomson of Merton.
[7]   It has been suggested by my colleague, Dr C. J. H. Watson, Fellow of Merton, that the Königsberg bridge problem could be avoided if one had a two-dimensional array of neurons "switching" from one path to another and back again rather rapidly, like the traffic flow in some American cities. But I do not see how one could arrange the "traffic lights" to change so as to route a message to its intended destination, unless there were an independent control of the traffic lights – which would require a third dimension to enable the "cables" controlling the lights to go under or high above the streets, in order to be out of the way of the traffic.

life but on those needed for any sort of communication of a causally coherent kind across a distance, separating two intelligences. It is like the transcendental deduction of the Lorentz transformations. We do not assume that people have bodies, but only that they have places. They are spatially located, and can interact with one another even though spatially separated; and this interaction takes a causally intelligible form. We apply Huyghens' principle, and argue that only in a three-dimensional space is high-fidelity signal transmission possible. It would be impossible to do Morse with a friend across a millpond – a two-dimensional medium – because the ripples "spread out", that is, go at different speeds, so that after only a short distance the 'message' would be blurred out of recognition and lost. And so in all spaces that are other than three-dimensional [8].

The exact details of the hi-fi argument, as I shall call it, are technical. We begin with the wave equation

$$\frac{1}{c^2} \frac{\partial^2 V}{\partial t^2} = \frac{\partial^2 V}{\partial x_1^2} + \frac{\partial^2 V}{\partial x_2^2} + \dots + \frac{\partial^2 V}{\partial x_n^2}$$

which imposes some conditions on the function U of $x_1, x_2, \dots x_n; t$. The wave equation has an underlying similarity with Laplace's equation (see p. 243). If we use the transformation $ict = x_{n+1}$ the wave equation becomes Laplace's equation. For example, by rewriting $x = x_1$, $y = x_2$, $z = x_3$, $ict = x_4$, the wave equation in (3+1)-dimensional space-time

$$\frac{1}{c^2} \frac{\partial^2 V}{\partial t^2} = \frac{\partial^2 V}{\partial x^2} + \frac{\partial^2 V}{\partial y^2} + \frac{\partial^2 V}{\partial z^2}$$

becomes Laplace's equation in four-dimensional space

$$0 = \frac{\partial^2 V}{\partial x_1^2} + \frac{\partial^2 V}{\partial x_2^2} + \frac{\partial^2 V}{\partial x_3^2} + \frac{\partial^2 V}{\partial x_4^2}.$$

The wave equation thus could be viewed as extending the egalitarianism of Laplace to apply over time as well as space – a sort of perpetual equality of opportunity. Less metaphorically, however, we should regard it as an equalizing condition, where the acceleration towards the average value is proportional to the distance away from it; such a condition is always overshooting the mark, and always oscillates around it instead of ever reaching it. Hence the waves.

We consider the "initial-value" problem of the wave equation. Only in three-, five-, seven- and other odd-dimensional spaces, but not in one- nor in any even-dimensional space does the wave equation satisfy Huygen's principle;

[8]   This argument is due to Richard Courant. See R. Courant in E. F. Beckenbach (ed.), *Modern Mathematics for the Engineer* (New York, 1956), p. 101; and R. Courant and D. Hilbert, *Methods of Mathematical Physics* (New York, 1962), vol. II, pp. 208–10, 688–91, 735–44, 763–6. It depends on the validity of Hadamard's conjecture, which we know not to be absolutely valid, but which Courant considers to be essentially correct.

that is to say, the message received at any point depends only on the initial conditions in a subspace of $n - 1$, or fewer, dimensions. For example, the light received at a point in a three-dimensional space depends only on what was being transmitted t seconds earlier on the two-dimensional surface of a sphere of radius ct centred on that point; any light from beyond that surface will not yet have reached the point in question; any from inside that surface will have already passed it. Similarly, a pulse of light going out from a point in a three-dimensional space will go out as an expanding sphere. After an interval t, it will reach every point on the surface of a sphere of radius ct, but no points outside, and will have passed all points inside. The wave point will be sharply defined, and eminently suitable for carrying messages. Whereas it would not have been in one-, two-, four- or any even-dimensional space.

Five-, seven-, nine-, eleven-, etc., dimensional spaces are still left in the field. Courant has an argument, which I have not been able to make my own, that the wave equations in these spaces do not admit as solutions relatively undistorted spherical waves of arbitrary shape, which is what we need for communication.

It may also be wondered whether the wave equation

$$\frac{1}{c^2}\frac{\partial^2 V}{\partial t^2} = \frac{\partial^2 V}{\partial x_1^2} + \frac{\partial^2 V}{\partial x_2^2} + \ldots + \frac{\partial^2 u}{\partial x_n^2}$$

is the only possible formula for the propagation of causal influence in space. We have not proved that a wave theory is the only possible alternative to a corpuscular one, and that if we do not have particles moving in what is otherwise a vacuum, we must fill our plenum with waves. But for purposes of communication, we can be sure that only a second-order differential equation will do, because with higher order equations, multiple diffraction would take place, so that high-fidelity reception would be impossible. J. Hadamard has conjectured that there are no equations essentially different from the wave equation that satisfy Huyghens' principle. If this conjecture is correct, we were right to want to fill the plenum with waves, and must furthermore be orthodox trinitarians about it, if we are to believe in there being more than one person, occupying different positions in space but nevertheless able to communicate with one another in a causally intelligible fashion.

There are many other considerations of a more technical and speculative nature which perhaps show that we need to have a three-dimensional space or, equivalently, a (3+1)-dimensional space-time. Thus Einstein's equations for empty space

$$R_{\mu\nu} = 0$$

imply flat space

$$R_{\mu\nu\lambda\rho} = 0$$

if the dimension of the space is less than four.

Again, the Dirac equation as normally used is valid only in four dimensions (of space-time), and Lanczos has noted that Maxwell's equations for free space, viz.

$$F^{\mu\nu}{}_{|\nu} = 0$$
$$*F^{\mu\nu}{}_{|\nu} = 0$$

have a symmetry that is revealed only in four dimensions [9]. Furthermore it happens that for $n = 4$, and only for $n = 4$, one can form a dual tensor from the Riemann tensor by the same operation used in forming the dual of $F_{\mu\nu}$.

More recently, Penney has argued that it is a necessary condition for the existence of a unified field theory embracing gravitation, electromagnetism and neutrinos that the world (i.e. space-time) be four-dimensional [10]. Leaving the neutrinos out of it, he is still able to show that only in a four-dimensional space-time do the two systems of Einstein's field equations (for gravitation) and Maxwell's equations (for electro-magnetism) determine their respective fields equally strongly; so that *if* a unified field theory exists which determines the phenomena of gravitation and electromagnetism as manifestations of the same phenomenon [11] and *if* the physical theories of gravitation and electromagnetism are correctly described by the Einstein and Maxwell theories [12], it follows that space-time must be four-dimensional.

They are big 'ifs'. As regards the first one, we have no warrant for saying that it is true, though plenty of reason, granted that it is true, for seeing why it should be. To base trinitarian conclusions on unitarian premises is both hallowed by tradition and intellectually pleasing. But it could yet prove the case that there was no unified field theory for both gravitation and electromagnetism – we might even come to see that it was necessary that these phenomena should be independent. For the present, a unified field theory is much more a *desideratum* than a firmly attested fact or well-established precondition of other deeply entrenched features of our thought.

As regards the second 'if' the case is reversed. The synthetic truth rating of Maxwell's equations is very soundly based indeed, and of Einstein's field equations reasonably so. But although they have considerable intellectual appeal for the applied mathematician and the working physicist, we have not yet been able to give an *a priori* exposition of their rationale. We have not shown that the most rational of all possible worlds is one in which there are electromagnetic and gravitational phenomena, which furthermore should satisfy equations discovered by Maxwell and Einstein. No doubt it will be done.

[9]   C. Lanczos, "The Splitting of the Riemann Tensor", *Reviews of Modern Physics*, XXXIV (1962), pp. 379–89.
[10]   R. Penney, "On the Dimensionality of the Real World", *Journal of Mathematical Physics*, VI, No. 11 (November, 1965), pp. 1607–11, from whom the preceding points are also taken.
[11]   ibid. Postulate II, p. 1611.
[12]   ibid. Postulate I, p. 1610.

But until it is, arguments from the unification of the two theories are founded on an *a posteriori* basis, rather than an *a priori* one.

Many other considerations have been adduced to explain the three-dimensionality of space [13], or the (3+1)-dimensionality of space-time (see, § 41, p. 194; § 43, p. 205), and much more needs to be done. Topology, the theory of groups and the theory of differential equations throw up many results showing differences between the small natural numbers, 1, 2, 3, 4, 5, etc.; and in some combination of these we may hope ultimately to achieve an Athanasian apology for the triune world we know so well.

[13]   The reader may care to refer also to A. S. Eddington, "Theory of Groups", in *New Pathways of Science* (Cambridge, 1935); reprinted in J. Newman, *The World of Mathematics* (New York, 1956), III, 1558; A. S. Eddington, *Fundamental Theory* (Cambridge, 1946), p. 124; A. S. Eddington, *Philosophy of Physical Science*, p. 169; E. A. Milne, *Proceedings of the Royal Society*, series A, CLX (1937), 8; H. Weyl, *Philosophy of Mathematics and Natural Science* (Princeton, 1949), p. 136; H. Weyl, *Das Kontinuum* (Leipzig, 1918), p. 67; and Aristotle, *De Caelo*, I, 1, 168a 9–29; G. J. Whitrow, *The Structure and Evolution of the Universe* (2nd ed. London, 1959), pp. 71–4.

V Return to time

§ 49
Time reversibility in classical physics

The laws of Newtonian mechanics and electromagnetic theory are indifferent to the direction of time. If at a certain instant we were to reverse all the velocities of all the constituents of a Newtonian system, the system would proceed to "unwind" and run backwards, and would be in exactly the same position at date t after the reversal as it had been at date $-t$ before reversal. Maxwell's equations can be solved with advanced, as well as retarded, potentials. If we were shown a film of a Newtonian system we could not tell whether the film was being run forwards or backwards. This contrasts, as we have seen, with films of human activities, biological processes or the phenomena of thermo-dynamics (see § 8, pp. 51–2). If we see men walking backwards, or plants growing smaller and contracting into seeds, or a cup of warm tea separating itself into hot raw tea and cold milk, we know that we are seeing the film the wrong way round. But if we saw the planets all going backwards in elliptical orbits, there would be nothing to indicate that anything was amiss. The laws of Newtonian mechanics would be satisfied as well by the one process as the other. There is no direction of time implicit in Newtonian mechanics, as there is in human activity, in biological process or in thermodynamics.

We need to distinguish what we are saying about Newtonian mechanics from two other claims, sometimes confused with it. We are claiming, first, more than that we can use Newtonian laws to retrodict as well as to predict. If we are given a functional dependence

$$U_1 = f_1(U_1, U_2, \ldots U_n; t)$$
$$U_2 = f_2'(U_1, U_2, \ldots U_n; t)$$
$$\ldots\ldots\ldots\ldots\ldots\ldots\ldots\ldots$$
$$U_n = f_n'(U_1, U_2, \ldots U_n; t)$$

by means of which we can predict the values $U_1', U_2', \ldots U_n'$ of the variables of a system, given their initial values, $U_1, U_2, \ldots U_n$, after an interval t has elapsed; we can then very often 'solve' the functional dependence, and find other functions, $g_1(\ ), g_2(\ ), \ldots g_n(\ )$, such that

$$U_1 = g_1(U_1', U_2', \ldots U_n'; t)$$
$$U_2 = g_2(U_1', U_2', \ldots U_n'; t)$$
$$\ldots\ldots\ldots\ldots\ldots\ldots\ldots\ldots$$
$$U_n = g_n(U_1', U_2', \ldots U_n'; t).$$

The condition that this should be so (namely that the "Jacobian" or functional determinant should not be zero) is often satisfied (see also § 36, pp. 174–5). But this does not mean that no direction of time is implicit in the functional dependence; for the functions $g_1(\ ), g_2(\ ), \ldots g_n(\ )$ are normally not the same as $f_1(\ ), f_2(\ ), \ldots f_n(\ )$. There is a difference then between predicting, where we use the functions $f_1(\ ), f_2(\ ), \ldots f_n(\ )$, and retrodicting, where we use the functions $g_1(\ ), g_2(\ ), \ldots g_n(\ )$. Of course, it might not be obvious which set of functions [$f_1(\ ), f_2(\ ), \ldots f_n(\ )$ or $g_1(\ ), g_2(\ ), \ldots g_n(\ )$] was to be used for prediction and which for retrodiction. This, however, is not to say that the functional dependence is indifferent as to the direction of time, but only that we do not see why it should go the way it does instead of the other way about [1]. In Newtonian mechanics not only can we retrodict, but the retrodicting functions, $g_1(\ ), g_2(\ ), \ldots g_n(\ )$, are the same as the predicting functions $f_1(\ ), f_2(\ ), \ldots f_n(\ )$. The form of the laws is the same in both directions. I have to be told whether a particular calculation is to be construed as a retrodiction from the condition $(x, \dot{x})$ [2] to a date $-t$, i.e. t *before* the condition, or as a prediction from the condition $(x, -\dot{x})$ to a date t, i.e. t *after* the condition. The calculation is the same.

We also, for complete caution's sake, need to distinguish the claim that classical physics is independent of the direction of time from the absurd claim

[1]   H. Reichenbach, in *The Direction of Time* (Berkeley, 1956), § 24, pp. 209–11, makes this confusion in arguing that "the problem of time direction is left by quantum mechanics in the same status as that in which it presented itself to classical physics" because "the sign on the right in Schrödinger's [time-dependent] equation can be tested observationally only if a direction of time has been previously defined."

[2] $(x, \dot{x})$ is short for $(x_1, x_2, \ldots x_n, \dot{x}_1, \dot{x}_2, \ldots \dot{x}_n)$ where $\dot{x}_1 = dx/dt$, etc. $(x_1, x_2, \ldots x_n, \dot{x}_1, \dot{x}_2, \ldots \dot{x}_n)$ are the $2n$ initial conditions in Newtonian mechanics, corresponding to the $(U_1, U_2, \ldots U_n)$ in the general case discussed in § 13, p. 75.

that for every set of initial conditions any retrodiction to any date $-t$ yields exactly the same results as a prediction from those initial conditions to a date $+t$. For this to be generally true, every variable would have to be of constant value, and nothing could ever change. We avoid the absurdity in Newtonian mechanics, because the initial conditions include the velocities of the constituents as well as their positions, and if we were to reverse the direction of time we should have to reverse the direction of each velocity, $\dot{x}_1, \dot{x}_2, \ldots \dot{x}_n$, since

$$\dot{x}_1 = \frac{dx_1}{dt} = -\frac{dx_1}{d(-t)} \text{ etc.}$$

In general, if a system were to start going backwards, it would develop *differently* from the way it develops going forwards. In this sense, it does make a difference which way it goes. But there is no difference between a system running backwards from initial conditions $(x, \dot{x})$ to date $-t$, or forwards from initial conditions $(x, -\dot{x})$ to $+t$. $(x, \dot{x}, -t)$ in Newtonian mechanics is the same as $(x, -\dot{x}, t)$. This is what we mean when we say Newtonian processes are reversible.

The reason why the laws of Newtonian mechanics are indifferent to the direction of time is that they can be formulated as second-order differential equations, in which the time variable occurs only in the form of second-order derivatives of other variables with respect to time. To take the simplest case, we say that

Force equals Mass × Acceleration
$$F = m\ddot{x}$$

where $\ddot{x} = d^2x/dt^2$. If we were to reverse the direction of time, we should have instead of d^2x/dt^2, the expression $d^2x/d(-t)^2$. But

$$\frac{d^2x}{d(-t)^2} = -\frac{d}{dt}\left(\frac{dx}{d(-t)}\right) = -\frac{d}{dt}\left(-\frac{dx}{dt}\right) = +\frac{d^2x}{dt^2}.$$

The two minuses, introduced by the two differentiations, cancel out. This is why second-order derivatives are indifferent to direction, and similarly natural laws expressed by a second-order differential equation. But whether a law of nature is to be expressed as a first-order or second-order differential equation depends on the number of variables used to express the law. In electromagnetic theory we may express Maxwell's equations either as *first*-order differential equations containing *six* unknown variables, or as *second*-order differential equations containing only *four* unknown variables. Newtonian mechanics, too, can be expressed in terms of different sets of variables. In more advanced treatments, we often deal with functions of position and momentum. Since, for constant mass, momentum is proportional to velocity we can consider this as a function of position and velocity – that is, of position and the first derivative of position with respect to time. We are particularly interested in laws expressed so far as possible in terms of spatial variables, and, if need be, their derivatives; for it is change of spatial parameters that represents change in its barest form.

By formulating a law in terms only of spatial variables, and, if need be, their derivatives, we are expressing it in the most general form possible. We therefore are content to consider Newtonian laws expressed in terms of spatial variables only, together with their derivatives; and we ask why in this form they are expressed by second-order differential equations, rather than, say, first-order ones.

If a law of nature could be expressed as a first-order differential equation involving only spatial variables, then it could be solved by a single integration, and would have only one arbitrary constant of integration for each spatial variable. It would mean that to calculate the state of a system we should need to be given only the value of each spatial variable at the outset. The relevant initial conditions are the spatial positions – the configuration – and nothing else. Once we had fixed these initial conditions, the state of the system at every subsequent date would be determined. If a law of nature is expressed as a second-order differential equation, then we need to integrate twice in order to solve it, and have to introduce *two* arbitrary constants of integration for each variable. In particular, if it is a second-order differential equation involving only spatial variables, then we need to be given the value both of each spatial variable and of its first time-differential, at the outset, if we are to be able to calculate subsequent states of the system. That is to say, we need both the initial configuration and the initial velocities as relevant initial conditions of the system. This means that our "phase space" is larger than our "change space" – that is, our "spatial space". We can count as different, in our phase space, systems we count as the same in our spatial space – e.g. two systems with the same configuration but different initial velocities. It is no longer true that every system that is the same as regards change space must develop in exactly the same way.

Is it important to have phase space larger than change space? I think it is. There are three arguments, none of them conclusive but each showing some intellectual inconvenience attendant on our having what I shall call, with some infidelity to fact and unfairness to Aristotle, Aristotelian mechanics, in which the natural motion of a system depends only on its spatial coordinates $(x_1, x_2, \ldots x_n)$.

Most positions in *phase* space are inaccessible to a system developing naturally, that is, without external interference, from any given initial condition (see § 15, p. 87). If change space embraces the whole of phase space, most of the changes it can describe can nevertheless be seen to be impossible. We can only avoid uneconomic redundancy of describable changes by having at least one variable in phase space an alteration of which does not itself constitute a change. Newtonian mechanics achieves it elegantly by having a change in the first-derivative of the variables of change space to provide the extra room for manoeuvre needed in phase space without themselves needing to be considered as part of change space.

Newtonian change space (i.e. spatial space) is assigned special features in order to discharge special responsibilities. We require it to provide us with a criterion of numerical identity. But unless our phase space is larger than change space, we should not be able to distinguish the conceptual framework which we identify things in, and need to be able to talk about them at all, from the causal theory we invoke to explain things. We should not have two independent concepts of identity and explanation. They would collapse into one undifferentiated concept. To identify and to explain would be the same thing, and we should be faced with Meno's problem of never being able to ask a question about anything, because if we could say what it was we were asking about we should know already what the answer to our question was [3]. If a thing is to go on being the same (numerically identical) thing in spite of having changed, it must in some respects have changed only continuously if at all. Else our criterion of numerical identity has gone, and we have no warrant for re-identifying the changed thing as being the same thing, although changed, as it was before it was changed. Therefore we must have as a criterion for numerical identity continuity with respect to at least part of change space. But if at least part, then either the whole, or else we have a division of change space into a fundamental or primary part, and an "accidental" or secondary part. Therefore, if we want to have all the variables of change space commensurable, all the dimensions on a par (see § 35, pp. 166–9), we must have, as the criterion of identity, continuity with respect to the whole of change space. But then if the whole of change space is bespoken to provide the criteria of continuity, we need to have phase space containing other variables as well, which can be invoked to explain, without making what we were talking about different from what it was.

In Newtonian mechanics the motion of an isolated single body depends only on its previous motion, not on its position: in Aristotelian mechanics it would depend not on its previous motion but only, if at all, on its place. Each body in a given position has a natural *nisus*, and if undisturbed each body will have a velocity of a magnitude and direction that would be purely a function of its position. In Aristotelian mechanics, therefore, we are faced with the alternatives either of abandoning position-indifference or of having an entirely static universe in which every position is as much a proper place for a thing as every other position, and therefore everything is in its proper place and nothing ever moves.

These three features of Aristotelian mechanics generate a metaphysics in which the world is a Spinozist substance, self-developing according to its own laws, and unable to develop otherwise than how its own being requires. Its future development must depend only on the positions of everything in the universe, and these cannot be different without the universe itself being a different one. Efficient causes collapse into formal causes and determinism

[3] *Meno*, 80d–e.

becomes a logical thesis, not a physical one as in the Newtonian philosophy. For the truth or falsehood of determinism even to be discussible, it must be possible to identify what we are talking about without having to give a complete causal state-description: which is to say that change space must be smaller than any putative phase space.

So far we have considered the world-view of a passive spectator. Normally, however, we view the world not as a single whole but as a number of relatively independent systems in which we – or other external agencies – may sometimes intervene. And the "forced" development of a system constrained by external intervention does not seem to be so very different under Aristotelian mechanics from what it would be under Newtonian. More effort, perhaps, would be required for effective intervention if velocity, rather than acceleration, was proportional to force. It would be more like stirring treacle than throwing balls. But life is often like that as it is, and anyhow there is no conceptual infelicity involved. By forceful intervention, we could make all regions of change space accessible to a system, no matter what its initial condition had been. Equally, we could make a thing become different from what it was, without making it be a different thing. In both these cases the possibility of intervention, instead of an enlargement of phase space, provides the added *Lebensraum* required for change space to be viable. The third difficulty – that of the isolated body – cannot be met directly: but it can be evaded, inasmuch as we never come upon a single isolated body, but only systems of interacting bodies. And for these it might well be that their motions depended only on their positions relative to one another – their configuration – not on their absolute positions. Difference of position between one body and another can be as relevant as difference of date between one state and another. All we need require is that their development be represented by a group of transformations that remained effectively the same if the position coordinates were themselves transformed by any transformation of the Euclidean group.

Even though intervention may be called upon to save Aristotelian mechanics from implausibility, it provides itself yet another argument for reversibility. For just as the function of Newtonian space was to give *things* room to change in, the function of Newtonian mechanics is to tell us how things change. And with things, if they change, they can change back again. Change is different from time. Not even the gods can make the past not have happened. They cannot *cancel* past events. But, where only things are concerned, they can undo them. It was to allow for the possibility of change that we introduced philosopher's space – but therefore also of reversible change. And equally our physics, which accounts for how things change, should account for their undergoing essentially reversible changes. When exception is taken to the temporal aimlessness of Newtonian mechanics, perhaps the ultimate answer is simply to put the cart before the horse and say that Newtonian mechanics is expressed in terms of second-order differential equations in order that they shall be reversible. Reversibility is, like homogeneity, a stipulation that we impose on a science of things (see § 14). It reveals a further facet of our concept of thingi-ness. Things are thought of as manipulable, subject to our wills. "Glory to Man in the highest", sang Swinburne, "for He is the master of things". Things, while still permanent, and in some ways resistant to our wills, are also the furniture of the universe, to be shoved around as we will. Hence the air of senselessness, and Berkeley's repugnance to Newton's and Locke's view of the world. Nothing seems to be leading anywhere – no purpose in its processes – everything just goes on and on and on – it all seems grey and dull and meaning-less. For this is how we wanted it. The prescription on which Newtonian physics was thought up was one in which things were subject to our wills and in which we could have reversible processes. It is therefore one which, apart from our wills, is value-free, one in which things seem to be leading nowhere, since they are available for us to go in any direction we want.

Not, of course, that Newton was an unknowing disciple of Swinburne or apostle of humanism. He takes a God's-eye view of creation, the atoms being created and given their primary qualities by *fiat* of God's will. But since the Newtonians, along with the voluntarists of the Middle Ages, insist on God's freedom to make any *fiats* He pleases, the effect, of depriving the created world of all value of its own, is the same. For the theist and atheist alike, the sense of temporal aimlessness in Newtonian mechanics remains. The celestial bodies and the minute corpuscles in their everlasting revolutions are not making any progress; they are not achieving anything; they are not getting any forrarder. There is no sense of purpose in the Newtonian world. There cannot be any teleology where end and beginning differ but in sign. And so we feel lost and ill at ease in an aimless world whose endless orbitings appear to lack all sense of direction. But the Newtonian world is of our own devising, to satisfy our own requirements. It is a world of things; and we require that they, like the good little things we want them to be, should exist independently of us, but should be recognizable by us and manipulable by us. The Newtonian world is the answer to these requirements, the best possible answer. Newtonian space provides the possibility of change in its simplest, most economical form. But it is unreasonable then to complain of the Newtonian system that it has no sense of temporal direction, for this is how we made it. It is a system of things, things we can change, things we change back again. Man, having made himself the master of things, cannot complain against them that they are not ends-in-themselves, and are indifferent to values. For that is just how we wanted them to be. If we put on grey-tinted spectacles, the world we see will seem to be grey.

§ 50
Time and probability

Time was reversible in Newtonian mechanics, partly because we wanted it so. But we could have it so only under certain conditions. Not only does Newtonian mechanics deal in only second-order differential equations, but it is a discrete and determinist system. Newtonian particles occupy exactly one point each, and no more. There are a definite integral number of parameters of any Newtonian system, and each has (ideally) an exact value, neither more nor less. And from them we can make a prediction (or retrodiction) which is 100 per cent true. If we relax these conditions, we lose the reversibility of time; particularly if we relax the last one, and deal with a continuum of probabilities instead of the two discrete truth-values, true and false.

Thermodynamics was the first branch of physics to become probabilistic – the Second Law of Thermodynamics is still the leading example of a physical law that is dependent on the direction of time. But thermodynamics has been beset by grave conceptual difficulties. These arise partly from historical circumstances, partly from an unclarity about the concept of probability itself. Historically, thermodynamics was grafted on to a Newtonian tree. The molecules of an ideal gas were thought of as Newtonian particles, obeying Newtonian laws, and perfectly elastic. Temperature and pressure were explained in terms of average kinetic energy, and average impulse. Perfect information was first assumed and then discarded in an averaging process. It was assumed to be possible in principle, and only disregarded because otiose. Thermodynamics was conceived of as Newtonian mechanics, redescribed more concisely, in a way more relevant to the practicalities of heat engineering. But the ghost of the hidden parameters remained unexorcised. Hypothetical cases could be conjured up in which a cup of tea would separate into hot tea and cold milk of its own accord, the molecules of water in the Red Sea divide by their own momenta to let the children of Israel pass. The direction of time seemed to be a function of ignorance. Sometimes, indeed, a Newtonian system must be moving from a "more probable" to a "less probable" state. In the fulness of time the whole universe would be doing this. When all secrets are revealed and all parameters known, probabilities will be resolved into certainties, thermodynamics will be absorbed back again by Newtonian mechanics, and we shall be back again with reversible time.

The influence of the Newtonian substructure has been the greater in that probability has often been viewed as a property of classes or ensembles, and classes are, by Russell's thesis of extensionality, equated with their members. Probability-talk then is, at one remove, talk about individuals, which, if the truth were fully told, could be characterized completely, without reference to probabilities at all. Probabilities are useful in calculating life-insurance tables, or conducting agricultural research, but they are not fundamental. They are a second best, in view of our limited information and limited time. Insurance companies would always prefer to know how long this particular proposer would live, rather than how long some one answering to the description of 'moderate smoker, moderate drinker, aged 34' would live. We randomize conditions only because we do not know, and have not time to find out, which of them are relevant and in what way. If we could isolate every factor without unacceptable expense of time, effort or money, we would. Probabilities, on this view, are partial descriptions of lots of individuals. They have their uses for finite fallible mortals, but could have no part in God's description of the world.

This view of probability is wrong. Probabilities should be regarded not as partial truths about lots of things, expressed compendiously though incompletely, but as natural generalizations of the concepts of truth and falsity. We generalize the discrete truth-values, true and false, into a continuous range of probability values, in much the same spirit as we go from integral numbers to real numbers, from series to integrals [1], from qualities to quantities, and from atoms-and-the-void to a smoothly filled plenum. Inasmuch as time is continuous, it is a peculiarly appropriate step: although, inasmuch as logic is

[1]   i.e. from $\sum_{r=0}^{n} f_r$ to $\int_0^n f(x)\,dx$. This has nothing to do with the integral numbers of the previous phrase.

discontinuous, we shall always have to retain discrete truth-values at some stage, and be ready to come down on one side of the fence or the other, and give a definite answer, Yes or No.

If probabilities are like truth values, they apply most naturally to propositions, but also to propositional functions. We can in symbolic logic say not only that p – a proposition – is true, but that $F(x)$ – a propositional function – is true. In ordinary symbolic logic, the latter is an unimportant usage, which can be dispensed with: in probability theory, however, it is of central importance, because it is only through propositional functions that we can apply the law of large numbers to make our estimates about probabilities come to face the facts. We are therefore concerned characteristically with propositional functions defined in some definite universe of discourse, and it is this that has led extensionally minded philosophers to regard probabilities as being proportions or frequencies of occurrence in some class or other.

The crucial difference between truth-values and probabilities is that the "law of contraposition" holds for the former but not for the latter. The proposition "If p then q" enables us not only to argue from the truth of p to the truth of q but from the falsity of q to the falsity of p. The nearest analogue to a conditional in probability theory, that given A the probability of B is α, cannot be similarly contraposed. If, by a series of tests, we come to the conclusion that the probability of B is something other than α, nothing follows. We cannot establish the falsity – or the truth – of A, by means of a conditional probability and further knowledge of B. Conditional probabilities, unlike conditional propositions, are one-way only. They are not symmetrical, either as regards inference, or, therefore, as regards time [2]. Given suitable (probabilistic) information, we can draw (probabilistic) conclusions and make (probabilistic) predictions. But we are not thereby enabled to make retrodictions, as we are when we are dealing with true or false predictions. With these, if any one description of the present state of affairs is true, every other description is false, and therefore every antecedent description from which one of these false descriptions could have been inferred must have been false too. And so, just as all except one of the possible descriptions is false, all except one of the possible descriptions at any given previous date must be false. Hence, by exhaustion, we can select the remaining one which must be true. Thus we can retrodict just as well as we can predict. Not so with probability.

This is not to say we cannot retrodict at all in probability theory. We may be able to. Granted B now, we may know the probability that A some time ago; in which case, we can retrodict. Only this will be an entirely different calculation from the probabilistic prediction of B, given A; and we may well have the one without the other: whereas with deterministic calculations they are but different sides of the same coin, and if we have the one we have the other too.

[2]   The argument is given more fully in J. R. Lucas, *The Concept of Probability* (Oxford, 1970), ch. XII.

Although probability theory permits retrodictions in principle, in practice it often does not. For a conditional probability is only applicable if the condition holds: it is no use knowing that given A the probability of B is α, if A is not in fact given. But to determine whether A is, or is not, given is to have a non-probabilistic Yes-or-No answer to the question "A or not?", and often, especially in quantum mechanics, merely to put the question is to introduce a disturbance which would make retrodiction pointless. It is therefore sub-stantially, although not entirely, true to say that inferences in probabilistic sciences are directed in time. The time-reversibility of classical physics was a special, not a typical, feature. It was due, at least in part, to the special role played by substances, things. Quantum mechanics has no comparable concept of substance; and therefore can be less resistant to the real nature of time.

§ 51
Time and modality

The introduction of probability has entirely altered the aspect of physics. Not only has it restored the need for time to have a direction, but it has opened afresh the question of how time is connected with modality. The connexion has been known much longer than that between time and probability, but has been subject to many of the same difficulties and confusions. The future is open, the present actual, the past unalterable. It is natural to try and view time as the passage from the possible to the necessary, and seek some such correlation as this:

Past	Present	Future
$\Box p$	p	$\Diamond p$

or, in the Polish notation,

Lp	p	Mp

But there are formidable difficulties. We have had positivist scruples in reifying modalities and have regarded objective possibilities with the same suspicion as objective probabilities. Moreover modal logic has until very recently been an obscure and opaque subject, quite incapable of casting any light on any field. Nevertheless we ought not to be deterred. Objective possibilities are in no worse a case than objective probabilities, and if our fundamental physics is irreducibly probabilistic, we shall have to swallow our positivist scruples anyhow. Some sorts of modal logic can be derived from the probability calculus [1], and we can view modal logic as a discrete analogue of the continuous spectrum of probabilities ranging between truth and falsehood. As such, possibilities are more appropriate than probabilities in all those cases where we are unable to assign numerical magnitudes [2], in particular in human affairs, where by our own choices we realize some possibilities and forgo others, and do irrevocable deeds. We are stressing the agency rather than the awareness of men. Time is not only the concomitant of consciousness, but the condition of choice and the passage from aspiration to achievement. And this is much more central to our concept of time. Although occasionally we contemplate in solitary stillness, by far the greater portion of our waking lives is occupied in doing and talking. Even thinking to oneself should often be accounted a type of activity, for in it we are seeking to achieve something – the solution of some problem – and as a result of it our thoughts are crystallized in a way they were not before.

In all these activities we are doing something. We are active agents, not bare sentient beings. We decide what we are going to do, and remember what we have done. Intention and memory are constitutive of our concept of ourselves as agents, and therefore of our concept of personal identity. I am the one person that I am because on certain issues it is up to me to decide what shall be done. In order to carry my decisions into effect I need at least a short-term memory, and the full concept of personality requires long-term memory too. I am responsible for what I have done, and I decide what I shall do in the knowledge that I shall thereafter be responsible for what I shall have done. I am not only an agent, but a rational agent, and reasons are omnitemporal (see § 13). In deciding what I shall do I am trying to make out what should – in some wide sense of 'should' – be done, and in answering for my actions afterwards I am adducing those same reasons which were, as far as I could make out, in favour of the course of action I actually undertook. Without long-term memory, there could be no responsibility. It is perhaps possible to think of oneself as an agent who would conscientiously address himself to the problem in hand; one might be anxious to do the best one could, but not under that description, for one would have no concept of having done anything, and so not of having done the best one could. All that one could envisage oneself wanting is that the best state of affairs should have come about; one could anticipate that one would regret it if the state of affairs that came about were less than optimum, but one could not even conceive that one would feel any peculiarly personal involvement in the situation or any more poignant sense of being in some way to blame for one's own failure to bring about the desired state of affairs. Some such

[1]   See Nicholas Rescher, "A Probabilistic Approach to Modal Logic", *Acta Philosophica Fennica*, XVI (1963), pp. 215–25.
[2]   See further below, § 53, pp. 284–7, for a sketch of how a probabilistic tense logic might be developed.

impersonal agency is perhaps conceivable, but it is not me. I am the man who I am because I not only do but have done, and because I not only aspire but aspire to have achieved. I know who I am because I can remember some of the things I have done, some of the places I have lived in, some of the people I have known, and because I can say what I am going to do and what some of my ambitions already are; and, fundamentally, it is only because I need not disown all my own actions that I am myself. Memory, as well as intention, is essential to the concept of personal identity that we actually have.

Memory is not all of a piece. Remembering how to talk, swim or bicycle is different from remembering who people are, which is different again from remembering what we have done, or that such and such is the case; even with what we have done, memory appears to differ with the lapse of time. A man who "loses his memory" loses his long-term memory of his actions, although in typical cases he retains both his memory of skills and a short-term memory that enables him to say what he was doing in the immediate past and to that extent remain a responsible agent. In concussion it is the other way about: the patient remembers his autobiography, but cannot remember what he was doing at, and immediately before, the time of the accident. This empirical difference in the psychology of memory appears to correspond with the conceptual difference between the memory required for responsibility and that required for personal identity. A responsible agent must be able to say, at the time of acting, what it is that he is doing, and this, as we have seen, requires memory – very good memory – of the immediate past. A person, if he is to be able to identify himself as a particular person, must have some memory, not necessarily a very good memory, of the relatively distant past; for a person is what he is only in virtue of what he has been, and, in particular, of what he has done and the intentions he has already formed.

Throughout life our attitudes and thought are coloured by our consciousness of the passage of time future into time past, and of the difference between them. The future is open: the young feel that life is full of possibilities; they can hope much, although also fear much, and can choose their aims, and seek to realize their ambitions. The past is fixed: it is irrevocable and definite; its verdict is final and cannot be unsaid; old men remember the deeds they have done, the deeds that made them the men they are. The young have promise, but have no achievement, no performance to their credit. The old have achieved something, but have done their performance, and cannot promise much more, or anything new.

The modal approach to time is much more consonant with our view of ourselves as responsible agents than is the tenuous time of classical physics. It has an inherent direction. It distinguishes one point, the present, from all others. And in its characterization of time as the passage from aspiration to achievement, it allows much more to personality and personal values than the dispassionate indifference to human concerns displayed by the rational theory of clocks. Equally, it requires those concepts of being able to do – or, in the passive form, being *feasible* – and of being able to remember (in the long-term sense) – or being *memorable* – to give content to the two modal operators required.

The topology is discrete. We do one thing after another, and it is not true, but absurd, that between any two decisions we take there is another intervening one. When we are talking in the first person it is quite correct to say what we did next, or what we did at the next moment. We see our lives as a succession of momentous decisions, and if this were our only approach to time we should not be entitled to maintain that time was continuous, or even dense. But since the argument from consciousness shows that time must be continuous (see §§ 6, 7), we should view the moments of decision as nodes, themselves discrete, set in a continuum of unmomentous instants. This is another illustration of the fact that no one approach will enable us to comprehend the whole of the concept of time (see §§ 1, 2, 12, 13, 15). The argument from consciousness, like the argument from change, secures the topological properties of denseness, continuity and linearity; the argument from agency is less convincing on these scores, but decisive on the equally important ones of the direction and the modality of time.

We can explain necessity in terms of possible worlds. A proposition is necessary if it is true in all possible worlds, possible if it is true in some (i.e. at least one) possible world, impossible if it is true in no possible world, and contingent if it is not true in all possible worlds. The great advance of recent years in modal logic has been due to the recognition that Leibniz's possible worlds can be taken relatively as well as absolutely [3]. In our particular case, we are concerned with what is feasible, and it is clear that what is feasible depends very much on what has already happened and what the present situation is: an enterprise may be feasible in one situation and not in another. We therefore extend our definition, and say that a proposition is *necessary in a given world* if it is true in *all worlds that are possible relative to that given world*; and similarly for *possible in a given world, impossible in a given world* and *contingent in a given world*. To avoid confusion, we shall add the modality *open* to apply to those propositions that are both contingent and possible, i.e. neither necessary nor impossible, and define a proposition as *open* in a given world if it is true in *some but not all* worlds that are possible relative to that given world.

With these definitions we can exhibit the differences between various modal logics as differences between types of relative possibility envisaged. In every case the relation of relative possibility must be reflexive, since a world must be possible relative to itself. If no other condition is satisfied, we have the modal

[3]   S. A. Kripke, "Semantic Analysis of Modal Logic, I: Normal Propositional Calculi", *Zeitschrift für mathematische Logik und Grundlagen der Mathematik*, VEB Deutscher Verlag der Wissenschaften, Berlin, IX (1963), pp. 67–96. The best exposition is given by G. E. Hughes and M. J. Cresswell, *An Introduction to Modal Logic* (London, 1968), ch. IV, esp. pp. 75–80.

system T. If the relation is symmetric, we have the Brouwerian system B; if transitive, S4; if transitive and symmetric – i.e. if it is an equivalence relation – S5. Each modal logic is thus correlated with some relational structure, a tree or a lattice, and the content of the modal operator corresponds to the topology of the relational structure. S5 corresponds to the degenerate case where the relation, being an equivalence relation, merely generates an equivalence class with no significant relational structure at all (see § 31, pp. 146–7). This would be the appropriate modality for Plotinus' eternity [4], or for the analyticity of modern logicians [5], but, just because it has no significant relational structure, it is inappropriate for temporal modalities. S4 is much more appropriate, corresponding as it does to a transitive relation, which will mirror, more or less, the transitivity of the temporal relations 'before' and 'after' [6]; but S4 fails to secure the uniqueness of the past. Although different courses of events could have resulted in end-states that were indistinguishable from one another, there would be at least this difference between them, that one course of events actually had occurred, and the others had not, although they might have done and it would not have made any significant difference if they had. According to Prior [7], some theologians and Marxists should find it unobjectionable that different paths should all lead to one and the same goal, and he quotes a philosophical justification from Lukasiewicz; but, as he points out, it runs counter to all our ordinary notions about time; and theologians, at least, are debarred from believing that the actuality of the past could ever fade or be expunged from the memory of God. Human memory may fail to provide a criterion of actuality for possible courses of past events, and even when we can remember past events, it does not follow that we can remember now all the things we could remember in time past. But at least sometimes we do, and our concept of memory would be incoherent unless our concept of the past were such that it was always logically possible for us to remember correctly our own correct rememberings. Moreover, if God exists, His memory provides a completely adequate criterion of what actually happened, and He will be able to remember not only external events that have happened but the fact that He was remembering them thereafter. We can express the uniqueness of the past by adding to S4 one further axiom:

$$Pp \,\&\, Pq \supset P(p \,\&\, q) \lor P(Pp \,\&\, q) \lor P(p \,\&\, Pq)$$

or, in Prior's Polish notation,

$$CKPpPqAPKpqAPKPpqPKpPq$$

[4]   *Enneads*, III, 7.
[5]   See E. J. Lemmon, "Is There Only One Correct System of Modal Logic?", *Proceedings of the Aristotelian Society, Supplementary Volume*, XXXIII (1959), § 6, pp. 35–7.
[6]   Only more or less: for 'before' and 'after' are asymmetrical, and therefore irreflexive, whereas relative possibility is reflexive. We can replace 'before' by the antisymmetrical 'not later than' and 'after' by 'not earlier than', but it is rather cumbersome.
[7]   A. N. Prior, *Past, Present and Future* (Oxford, 1967), pp. 28–9.

which can be interpreted either theologically or logically. In the theological interpretation we understand the operator P as 'God can remember'. If God can remember p and also can remember q, then either He can remember p and q – if p and q were contemporaneous – or He can remember q and His then at the same time remembering p – if p was before q – or He can remember p and His then at the same time remembering q. In the logical interpretation we interpret P as a past-tense operator, and assert that, if it both was the case that p and was the case that q, then either it was the case that both p and q or it was the case that both q and at that time it was the case that p or it was the case that both p and at that time it was the case that q. Either interpretation is intuitively acceptable. If we add the appropriate axiom to S4 we obtain the system discovered by Dummett and Lemmon, S4.3 [8].

The additional axiom can be expressed as

$$L(Lp \supset Lq) \lor L(Lq \supset Lp)$$

$$ALCLpLqLCLqLp$$

or, equivalently,

$$L(Lp \supset q) \lor L(Lq \supset p)$$

$$ALCLpqLCLqp$$

or, again equivalently,

$$Mp \,\&\, Mq \supset M(Mp \,\&\, q) \lor M(p \,\&\, Mq)$$

$$CKMpMqAMKMpqMKpMq.$$

It is the last form that is closest to Prior's tense logic, with P ("it was at some time the case that") being analogous to M ("it is in some world the case that"). We should notice that the S4.3 axiom does not have the disjunct $M(p\&q)$. This is because it is redundant, since under the other axioms of modal logic $p \supset Mp$, and hence $M(p\&q) \supset M(Mp\&q)$. In our interpretation of modal logic in terms of possible worlds, we specified that the relation of relative possibility must be reflexive, since a world must be possible relative to itself. In our normal use of the word 'past', however, we do not understand it to include the present. We must therefore either add an extra disjunct $P(p\&q)$ to the consequent of the special axiom of S4.3, as Prior does, or understand P to mean not exclusively the past, but the past-or-present. We shall do the latter. It will keep us nearer modal logic. We leave the question of density on one side, and take the present as the last instant of the past, and, where need be, the first of the future; being, as the Greeks thought, the connecting link between the two [9].

[8] M. A. E. Dummett and E. J. Lemmon, "Modal Logics between S4 and S5", *Zeitschrift für mathematische Logic und Grundlagen der Mathematik*, VEB Deutscher Verlag der Wissenschaften, Berlin, III (1959), p. 252.
[9] *Physics*, IV, 13, 222a 10.

The difference between S4.3 and S4 is that the corresponding relation for S4.3 is linear while that for S4 is not. S4 can be represented as a tree

or a semi-lattice

or a lattice

or a line

but the additional thesis of S4.3

$$Pp \,\&\, Pq \supset P(p \,\&\, q) \vee P(Pp \,\&\, q) \vee P(p \,\&\, Pq)$$
$$CKPpAPKpqAPKPpqPKpPq$$

rules out all these except the line. For if any "possible world" or state of affairs (represented by a node on the diagrams) is such that p is true of it, then every possible world or state of affairs to which it is related or from which it is accessible (represented by a lower node in the diagrams) is such that Mp is true of it too. It is immediately clear in the case of the tree and the semi-lattice that if p were true of the left node and q were true of the top right node, but not vice versa, then Pp and Pq would be true at the bottom node, without either $P(p\&q)$ or $P(Pp\&q)$ or $P(p\&Pq)$ being true there. In the case of the lattice we need to consider the two nodes in the line below the top; and again if p were true of the left one and q of the right, but not vice versa, then Pp and Pq would be true at the bottom node (and three others as well), but neither $P(p\&q)$ nor $P(Pp\&q)$ nor $P(p\&Pq)$. Only in the linear case can we be sure that if any at node $Pp\&Pq$ is true, then since it must be because there is some node above it (or to be exact, not below it) at which p is true and some node at which q is true, and since there is no branching, either the two nodes are the same (in which case $P(p\&q)$ is true), or the one at which q is true is lower than the one at which p is true (in which case at that node $Pp\&q$ is true, and therefore $P(Pp\&q)$ is true at all lower ones) or the one at which p is true is lower than the one at which q is true (in which case at the node at which p is true, Pq is true too, and so $p\&Pq$ is true, and $P(p\&Pq)$ is true at all lower ones). Thus S4.3 is linear, whereas S4 need not be, and is the correct modality for the past.

No argument from memory shows that the future must be represented by S4.3, and the fact that it is characteristically alterable counts against it. Although some future events are already inevitable, others are not, but are what Aristotle calls τὰ ἐνδεχόμενα ἄλλως ἔχειν [10], capable of being otherwise. The distinction is difficult to draw exactly. Even eclipses, Aristotle's best example of the inevitable, might be averted by sufficiently drastic rocketry. In general, it depends on the agent: we need to specify "avertible by whom". There are many political events that are inevitable so far as I am concerned, but are entirely within the capacity of Parliament or Congress to alter. These ambiguities notwithstanding, it is true of every agent that he can choose within some range of alternatives what he will actually do. And, as the French word 'actuel' suggests, it is this act of realizing one possibility rather than others that constitutes the present in contrast to the future. The future, therefore, is not unique. There are many possible worlds, or possible states of affairs, that are feasible for the agent in question. We take feasibility as the typical characteristic of futurity: that is to say, if we define *worlds that are possible relative to a given world* as *feasible worlds*, then the relation will be one of futurity.

It may be helpful to draw an analogy with family trees. If we consider only the male sex (in order to avoid non-branching backwards), we notice that whereas every man has one and only one father, he may have more than one son. Each man has a unique line of paternal ancestors, but may have many lines of male descendants. Each man's own family tree will thus be a single line so far as his ancestors are concerned, but a set of branching lines, as far as his descendants are concerned. In the usual representation it is an upside-down tree. However, in our representation of both past and future, we shall follow Prior [11], and put our diagrams on their sides, with the past towards the left

[10] *Nicomachean Ethics*, VI, 6, 1141a 1.
[11] A. N. Prior, *Past, Present and Future* (Oxford, 1967), p. 127.

and the future towards the right, with the line therefore extending linearly towards the left and branching towards the right:

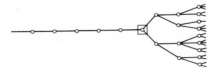

In such a diagram there is a particular point, the first node at which branching occurs (marked by circumscribed square), which represents the present, the point of view from which the past is past and the future future. It is clear that the appearance of the tree depends on which point is taken as the present. If the point of view had been four nodes earlier, the tree would have appeared:

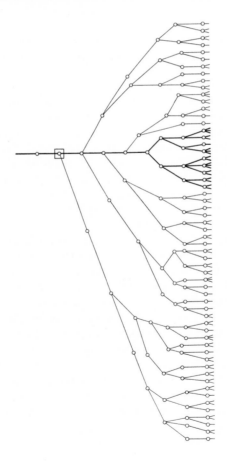

If, on the other hand, the point of view had been two nodes later, the tree would have appeared:

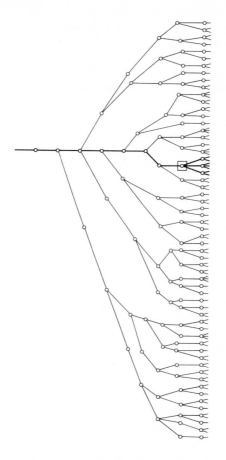

In this last case we show faintly all the might-have-beens. The faint branches were feasible at an earlier date – the date of the second tree – but now no longer are. At every node a choice has to be made, and the possibilities not chosen cease to be real possibilities. If we revert to the vertical diagram, we have a tree that is continually shedding its branches, leaving a bare trunk up to the point representing the present, and branching only above it.

It is difficult to capture this intuitive characterization of the future by means of modal logic. For one thing, although S4 allows a tree-like structure, it does not require it. Clearly, it is compatible with S4.3, and could be satisfied by a linear non-branching future. If we adopt an S4 modality for the future operator, we allow that the future may be open, but do not insist that it shall. But that is as it should be, for the future is not always open. Often it seems, and perhaps sometimes correctly, that we have no alternative. And various philosophies of determinism are intelligible, if repugnant, and cannot be ruled out on grounds of logic alone. All we can require from logic is that the openness of the future should not be foreclosed by our logic: and S4 is on that score all that we could ask.

Other difficulties arise from the interpretation of the modal operators and the nodes. The nodes represent worlds not dates. When we say that there are many possible futures we mean that there are many future courses of action that are feasible, not that there is some non-linear ordering of future dates. My going to the library tomorrow and my not going to the library tomorrow are both open possibilities, and there are two possible future worlds, in one of which I go to the library and in the other of which I do not: but there is only one tomorrow. If we adopt the tree and lattice models of temporal relations, or talk of the past being linear and the future branching, we must be careful to keep the distinction between worlds and events very clear, the more so since in ordinary language either can be referred to by the adjectives 'past' and 'future', and in philosophical contexts the extreme tenuousness of time, as I am expounding it, may lead people to give it body by identifying it with states of affairs or possible worlds (see, e.g., § 7, pp. 37–8; also § 2, pp. 9–13, and § 15, pp. 89–91). We need to emphasize the distinction between the different nodes on different branches representing possible worlds or states of affairs, and their projection on the time-axis, which represents their dates, which in turn must be linear, even if the possible worlds are not.

Modal logic enables us to formulate in logical terms the difference between the future and the past, and hence pave the way to a more adequate character-ization of time than the tenuous space-like entity generated by the rational theory of clocks. But it is dangerous to embark on a radical reconstruction of tense logic in purely modal terms. Although our tenses reflect our fundamental view of ourselves as agents, to whom therefore modal operators are relevant, they are influenced by other considerations as well. We need therefore to examine our actual use of tenses, rather than lay down a new tense logic of our own.

§ 52
Tenses

Verbs conjugate not only in tenses, but in persons and moods too. It is no accident. The covert egocentricity of our tenses parallels the overt egocentricity of the first and second persons, because my attitude to the future and the past is primarily a personal one, just as my attitude to myself and to you is. The auxiliary verbs used to express the future often are basically modal verbs – 'shall', '*sollen*', 'should'; even the peculiar 'I shall', 'you will', 'he will' can be explained as a natural interplay of moods and persons devised to convey some force of futurity. But just as there are redundancies and difficulties built into the first, second and third persons of verbs, there are redundancies and difficulties in our use of tenses that are peculiarly resistant to clarification.

Tenses are covertly egocentric because the force of a tensed sentence depends on the date at which it is uttered. As a first approximation we can carry over the analysis of the adjectives 'present', 'future' and 'past' given in § 4 to the comparable tenses, and say that the present tense is appropriate for talking about events either contemporaneous with the instant at which I am speaking or occurring within some indefinite interval, depending on context, which includes the instant at which I am speaking. The future tense is appropriate for events occurring after the present instant, the past for those before. Tenses resemble persons. It is a rule of language that if I want to tell you that J. R.

Lucas is ill, I use the words 'I am ill', whereas if you want to tell me, you use the words 'You are ill', and if you want to tell some one else, you use the words 'He is ill'. This rule is not entirely a rule whereby references can be identified, although in many cases it enables us to identify the person referred to. But even if identification is secured by other means – if I say 'I, J. R. Lucas, ...' – it is still necessary to use the correct part of the verb, which indicates not merely the person referred to, but his relation to the person speaking or the person spoken to. Similarly, even if I identify the date of an event I am describing, I am still bound to use the correct tense, and if you on some other occasion are describing the same event, you must use the tense appropriate for that occasion, just as you must change the person of the verb, replacing my 'I' by your 'you', and vice versa. If I say 'I am ill' and you thereupon want to contradict me, you have to say 'You are not ill'; and if at some later date I want to contradict myself, I have to say 'I was not ill' just as you would have to say 'You were not ill'. To this extent tenses, although important stylistically, are not fundamental to logic. The same proposition will be expressed now in one tense, now in another, just as it will be expressed now in one person now in another. But what we, as logicians, are concerned with is not the different expressions of the same proposition, but the features that all the different locutions have in common. Whether you say to me 'You, Lucas, are ill', or to somebody else 'He, Lucas, is ill', does not matter: what matters is your claim that Lucas is ill, which may or may not be true and which will be consistent with some other propositions and inconsistent with others.

In this important sense, the tense of an utterance is not part of the meaning of the utterance, any more than the person is. When you say to me 'You are not ill', I can contradict or agree with you without using the second person at all, and when you say 'You are not ill', I may later contradict or agree with you, without using the present tense at all. Nevertheless, in another sense, the tense and the person of the verb affects the meaning of the utterance in an important way. I mean something very different If I say 'I am ill' from what I mean if I say 'You are ill' or 'I was ill'. Given a particular context, a particular person uttering a sentence on a particular occasion, the token-reflexive terms convey information and must be accounted part of the meaning of the sentence. It depends on our interest. If our interest is primarily textual, if, that is, we are concerned with the *ipsissima verba* of an author actually used by him in a given context, then token-reflexive terms are of central importance to the meaning. If, however, our interest is Platonist, if, that is, we are concerned not so much with what was actually said in one context as with what could have equivalently been said by other contributors to the discussion in other contexts, then we shall construe the token-reflexive terms as varying conversely with the various contexts in order to secure an invariant content, which alone is the focus of our concern; we shall "see through" the token-reflexives to what each one, in each particular context, refers to, and concentrate on that, discounting the variability of the language used in varying contexts as an adventitious infirmity of human language. Our aim will be to consider all logically equivalent locutions together, in order to discuss whether the equivalence class constitutes a true or a false proposition; and the test of equivalence is based on that of contradiction with the aid of double negation. As Platonists we are not concerned with tenses, and disregard, so far as we can, the tense structure of the locutions actually used to express propositions. Even if we cannot have a language totally free of token-reflexive terms, we can reasonably reckon to "see through" them from any particular passage whose truth or logical import is under discussion. And in this sense the main tense of an utterance does not constitute part of its meaning.

Reichenbach has developed a sophisticated account of tensed discourse, which allows for the ways in which the tense of an utterance does, and does not, affect its meaning [1]. He distinguishes not only the date of the event referred to and the date of speaking, but a third date, "the point of reference" of the discourse. "The visitors said they had eaten on the way" – the pluperfect enables us to date their meal not only before my speaking to you of it, but before their speaking to me of it, which I am only indirectly speaking of to you. Reichenbach symbolizes the pluperfect

$$\overset{\longmapsto\quad\quad\quad\quad\longrightarrow}{\underset{\text{E}\qquad\quad\text{R}\qquad\quad\text{S}}{}}$$

where E is the date of the event, R the "point of reference" and S the date of speaking. It is clear that a similar analysis applies to the "plufuture" (which he names the "posterior future")

$$\overset{\longmapsto\quad\quad\quad\quad\longrightarrow}{\underset{\text{S}\qquad\quad\text{R}\qquad\quad\text{E}}{}}$$

and to the various possibilities for the future perfect

$$\overset{\longrightarrow}{\underset{\text{S}\quad\text{E}\quad\text{R}}{}}\ \text{or}\ \overset{\longrightarrow}{\underset{\text{S,E}\quad\text{R}}{}}\ \text{or}\ \overset{\longrightarrow}{\underset{\text{E}\quad\text{S}\quad\text{R}}{}}$$

and similarly to the corresponding past-future tenses, which have no special name but are expressed by the form '... was going to ...':

$$\overset{\longrightarrow}{\underset{\text{R}\quad\text{E}\quad\text{S}}{}}\ \text{or}\ \overset{\longrightarrow}{\underset{\text{R}\quad\text{S,E}}{}}\ \text{or}\ \overset{\longrightarrow}{\underset{\text{R}\quad\text{S}\quad\text{E}}{}}\ .$$

Reichenbach also applies his analysis to the perfect tense, which he symbolizes

$$\overset{\longrightarrow}{\underset{\text{E}\quad\text{S,R}}{}}$$

for when I say 'I have arrived', I am referring to my present state consequent upon the past event of my arrival, in contrast to the simple aorist, which he symbolizes

$$\overset{\longrightarrow}{\underset{\text{E,R}\quad\text{S}}{}}$$

because the past event is then being described from a past point of view.

[1]• Hans Reichenbach, *Elements of Symbolic Logic* (New York, 1951), § 51, pp. 287–98.

The virtue of Reichenbach's analysis is that it enables us to distinguish the unvarying temporal relation between the reference point and the date of the event from the essentially variable relation between them and the date of utterance. Tensed utterances, like those containing personal pronouns or explicitly token-reflexive words such as 'now' or 'here', depend on their context of utterance. The utterance of the words 'Peter went to Cambridge' may be true if uttered on one occasion, false if uttered on another, just as the utterance 'I am over six foot tall' is true if uttered by J. R. Lucas, but would be false if it were uttered by Princess Margaret. We have a systematic way of correlating utterances of the same and different sentences by the same and different people so that the same thing – the same proposition or the same statement – may be affirmed or denied by various people. 'I am over six foot tall' in J. R. Lucas's mouth says the same thing as 'You are over six foot tall' in the mouth of some one else addressing J. R. Lucas, or 'J. R. Lucas is over six foot tall' in the mouth of a third party, or 'He is over six foot tall' when the context makes it clear that 'He' refers to J. R. Lucas. In the same way 'There will be trouble in Ruritania in 1911' uttered before 1911 and 'There was trouble in Ruritania in 1911' uttered after 1911 say the same thing. The test is contradiction. If two utterances in their respective contexts of utterance contradict each other then the contexts are such that they are affirming and denying the same proposition; and similar utterances, only with one of them suitably negated, would have expressed the same proposition. It is characteristic of such utterances which express the same proposition, that the question whether one should use the present, past or future tense depends on whether the date at which one is speaking is contemporary with, after or before the date one is talking of: and this will vary systematically with actual occasions of utterance. But whether one should use the simple present, past or future, or certain more complicated variants such as the perfect, pluperfect and future perfect will depend on features of one's discourse that are independent of the date of utterance. If ever it was right to use the future perfect – "By the time he gets their letters, he will have made up his mind" – it will be correct on certain other occasions to use the perfect or the pluperfect – "Now that he is getting the letter, he has made up his mind" or "By the time he got the letter, he had made up his mind" – but never the aorist, say, or the past future "When he got the letter he was about to make up his mind". The fact that the date of his making up his mind precedes the date of his receiving the letter is one that does not alter with the occasion of utterance, and although our tenses vary with the occasion of utterance, they do so in a systematic way, in order to preserve the underlying temporal invariances too.

Reichenbach's analysis has been further refined by Kenny, who takes into account the distinction between instants and intervals, and a further distinction between those intervals with definite end-points and those specified only indefinitely [2]. Kenny is able to explicate the different forces of different sorts

[2]   Anthony Kenny, *Action, Emotion and Will* (London, 1963), ch. 8, pp. 173–5.

of verb – for instance 'I have loved her for seven years', which implies that I still love her now, in contrast to 'I have built a house', which implies that I am not still building it now. These are valuable distinctions for the elucidation of the different types of performance and activity that a person may undertake, and may well throw further light on our understanding of time as the condition of activity. Reichenbach's analysis can also be extended to more complicated cases, such as 'I shall have been going to see John', which might be represented

by
$$\text{S} \qquad R_2 \qquad \text{E} \qquad R_1 \longrightarrow$$
[3]

or, perhaps, by
$$\text{S} \qquad R_2 \qquad R_1 \qquad \text{E} \longrightarrow$$

Although we can iterate tenses indefinitely, the resulting locutions begin to suffer from such great indefiniteness of reference – is E before or after R_1? – that it becomes unworkable unless we can supplement our tenses by definite dates (see § 53).

The use of tenses is an affront to logicians. They lack a sense of time. Their study is of timeless relations that hold between timeless propositions. But since it is also a study of the most general features, characteristic of all discourse, including that about changing events, they are often tempted to freeze the variable flux of temporal phenomena into the rigid immobility of Platonic truth, arguing either with Plato, Leibniz, Bradley, McTaggart and the poets that time is unreal, or with Pythagoras, Iamblichus and the fatalists that time is not really time, but is and always has been already fixed. Logicians are natural Platonists, because they always hope to overcome the original sin of their own utterances' egocentricity, and entertain propositions whose meaning and truth is entirely independent of context. But once we start talking timelessly about time, we make ourselves susceptible to a philosophical schizophrenia: on the one hand we seek to purge our discourse of all terms whose meaning depends on the occasion of utterance, terms such as 'now', 'then', 'past', 'present' and 'future', and tensed verbs; on the other hand we feel we cannot do justice to our concept of time unless we refer somewhere to the fact that it is something we experience, and that the present is radically different from the past and the future, which are themselves unlike each other. We may be led to speak so timelessly about time that it ceases to be about time at all, and then to reintroduce on a new metaphysical level the time we had taken such pains to expel. Pseudo-Archytas, who may have lived some time between 200 B.C. and A.D. 200, may have been the first philosopher to suffer in this way. Iamblichus, the Neoplatonist, made it a central tenet of his metaphysical system that there were two kinds of time: intellectual time, which contains the relations of before and after and is indivisible, permanent and stable, and sensible time, which is changing, fleeting and unreal. Intellectual time, according to Damascius, is like the waters of a river, considered throughout its course, which all bear a stable spatial relationship to one another, some being

[3]   A. N. Prior, *Past, Present and Future* (Oxford, 1967), p. 13.

higher, others lower, even though they all are flowing down the river's course; whereas sensible time is like the waters of a river that flow past one point, which are always coming and passing on, their place being taken by fresh water. It is a proposition of intellectual time that the Trojan war happened before the Peloponnesian war, and the truth of this proposition is stable and the temporal relationship of the Trojan and Peloponnesian wars will never change: but it is only a matter of sensible time that these events are now past, and were once present, and before that were future [4]. McTaggart is similarly torn. Positions in time, he says, are distinguished in two ways: either by the relation of "earlier than" (or "later than") or by being either past or present or future [5]. The former he calls the B series, the latter the A series:

> Since distinctions of the first class are permanent, it might be thought that they were more objective, and more essential to the nature of time, than those of the second class. I believe, however, that this would be a mistake, and that the distinction of past, present, and future is as *essential* to time as the distinction of earlier and later, while in a certain sense it may . . . be regarded as more *fundamental* than the distinction of earlier and later. And it is because the distinctions of past, present and future seem to me to be essential for time, that I regard time as unreal. [6]

For we cannot predicate the terms 'past', 'present' or 'future' tenselessly of temporal instants or temporal events. We cannot say tenselessly that M is present, past and future, but only that it *is* present, *will be* past, and *has been* future, or that it *is* past, and *has been* future and present, or again that it *is* future, and *will be* present and past. But, asks McTaggart, what is meant by "has been", "will be" and the tensed "is"? Tenses can be explicated only in terms of past, present and future, and therefore we must either proceed down an infinite regress, or allow that every temporal instant is both past, present and future, which is a patent contradiction [7].

McTaggart's reasoning can be faulted at both these points [8]. Once we recognize the contextual dependence of the terms 'past', 'present' and 'future' and can explicate them as meaning 'before the occasion of utterance', '(roughly) simultaneous with the occasion of utterance', and 'after the occasion

---

[4]   S. Sambursky, "The Concept of Time in Late Neopolatonism", *Proceedings of the Israel Academy of Sciences and Humanities*, II, No. 8 (1966).
[5]   J. M. E. McTaggart, *The Nature of Existence* (Cambridge, 1927), ch. 32, §§ 305–6, vol. II, pp. 9–10. See also "The Unreality of Time", *Mind*, XVII (1908).
[6]   McTaggart, *The Nature of Existence*, § 306, p. 10.
[7]   ibid. §§ 330–2, pp. 21–22.
[8]   See, e.g., D. F. Pears, "Time, Truth and Inference", *Proceedings of the Aristotelian Society*, LI (1950–1), pp. 1–9; but see further M. A. E. Dummett, "A Defence of McTaggart's Proof of the Unreality of Time", *Philosophical Review*, LXIX (1960), pp. 497–505, esp. pp. 502–3; and Richard M. Gale, "McTaggart's Analysis of Time", *American Philosophical Quarterly*, III (1966), pp. 145–52.

of utterance', then the contradiction disappears. The fact that events and instants admit of all three terms is no more surprising than that the same place can be referred to on different occasions or by different persons as here or there, or that the words 'I' and 'you' can both refer to the same person and each to different ones. McTaggart is making the same mistake as Plato, who extracted a contradiction from the fact that one and the same person or finger can be called big (in comparison with smaller ones) and small (in comparison with bigger ones) or that a beautiful ape would still be very ugly, regarded as a human being [9]. But 'big' and 'small' are not absolute terms. Nor are 'past', 'present' and 'future'. It is true that every event and every temporal instant is (tenselessly) past, present and future, but *from different points of view*. It is no more a contradiction than Ramsey's "I went to Grantchester yesterday" – "Oh, did you? I didn't." [10]

Equally, the infinite regress is one down which we neither need regret being able to go, nor are obliged to go. It is true, as we shall see in the next section, that we can develop tense logic to accommodate as many iterations of tense operators as we please: there is no objection, save that of tedium, to our doing so. But neither on us nor on McTaggart is there any call to do so. It was only because McTaggart thought he ought to talk tenselessly about time that he could not rest content with the acceptable conclusion that *M* either is present, will be past and has been future, or is past and has been future and present, or is future and will be present and past.

Although we can fault McTaggart's arguments, his predicament is a real one. He feels the pull of Platonism, but senses that time is essentially unplatonic. It is partly a general epistemological difficulty about Platonism, but in the case of time it is reinforced by firmly entrenched linguistic habits, which reflect the peculiar connexion of time with persons, both as conscious beings and, more particularly, as rational agents.

Aristotle objected to Plato's theory of forms on the ground that they were not connected to our actual experience [11], and we make a similar objection to those modern Platonists, such as Quine, who seek to purge language of all personal pronouns, token-reflexives and tenses [12]. Although we can go along with them to a very large extent, in the interests of scientific objectivity, we cannot follow them all the way, because a language in which there are no token-reflexive words has no anchorage in experience. A map set up in a street in a strange town needs to have an arrow saying 'YOU ARE HERE' to enable the visitor to locate himself. Many statements might be made about J. R. Lucas, describing and characterizing him fairly fully. But at some stage I must be able to say 'I am J. R. Lucas' and identify myself to my hearers as the person being

[9]   *Phaedo*, 102b; *Republic*, VII, 523c–524d; *Hippias Major*, 390a.
[10]   F. P. Ramsey, *Foundations of Mathematics* (London, 1931), p. 289.
[11]   *Nicomachean Ethics*, I, 6, 1096a 11–1097a 14.
[12]   W. V. Quine, *Word and Object* (New York, 1960), pp. 193 f., 226 f.; or *Elementary Logic* (New York, 1965), p. 6.

talked about. Otherwise, although they can talk about J. R. Lucas in a grammatically correct fashion, they will never know who, really, is being talked about, nor will they be able to assess statements made about him – only I am able to speak with first-hand authority about my intentions or sensations. So, too, we cannot do without the words 'this' and 'that', and so too we cannot speak temporally unless we have at our disposal either tenses or words or phrases like 'now' or 'the present instant'.

We cannot achieve complete non-egocentricity in our characterization either of time or of space. We can characterize many of the formal properties, but in each case we sense there is something left out if we are too adamant in excluding ourselves. It is not to any entity X outside the A series that each of its members must bear a relation either of pastness, or of presentness, or of futurity [13], but to a particular temporal point of reference, provided by McTaggart himself or some other conscious being. We can detach our consideration from any one particular temporal viewpoint; but what we are considering ceases to be time if we do not allow that it is characteristically viewed from some viewpoint or other. Similarly with space. There is no one frame of reference which it is necessary that we should adopt: but it is necessary that we should adopt some one frame of reference or other (see § 30, pp. 143–5) Neither with time nor space can we achieve the ideal of non-egocentricity absolutely. But whereas with space, the egocentricity could in principle be limited to my choice of a frame of reference, with time the egocentricity is more insistent, in that I must be able to correlate the frame of reference with the occasion at which I am discussing it (see § 7, pp. 36–7, and § 9, pp. 58–9). It would be odd, but not inconceivable for me to refer to a space, and then admit under questioning that my own location could not be referred to in that frame, since I was not a spatially located being at all; but it would be unintelligible for me to offer a frame of temporal reference within which I could not refer to the date at which I was then speaking. It is part of the concept of time that it is connected to us whereas it is not absolutely necessary either that space should be connected to us (see § 28, pp. 133–4) or that a conscious being should be located in space (see § 2, p. 7; § 44, pp. 216–17; § 55, p. 304). A God's-eye view of space is conceivable, but not, as it has been traditionally understood, a God's-eye, totally impersonal, and in that sense non-egocentric, view of time.

The essential egocentricity of time is reflected in the ineliminability of tenses, which we may compare and contrast with the eliminability of the first and second persons. If we are attempting to carry out a Platonist programme of eliminating all personal pronouns and other token-reflexive terms, we feel a linguistic awkwardness in replacing 'I' and 'you' by proper names. If someone heard me, J. R. Lucas, uttering the words 'J. R. Lucas is over six foot tall' he would assume that I, the speaker, was not J. R. Lucas, and on finding that I was might feel that I had been somewhat disingenuous, unless either he

[13] McTaggart, op. cit., § 328, p. 20.

realized I was a philosopher talking about token-reflexives or else he was a monarch who regarded it as *lèse-majesté* for the first person to be used in his presence of any one but himself. But although awkward, and sometimes misleading, we should hesitate to say that failure to use the first or second person was positively wrong. But it is positively wrong to use a past tense to refer to an event that is still to come, or to use a future of what has already happened. I cannot now say that Caesar will be killed in 43 B.C., in the way that I can say 'I, J. R. Lucas, am over six foot tall'. We have no system of tenseless sentences for referring to dated events, although Quine has yearned for them and called them eternal sentences. We might adopt such locutions on Platonic principle, but linguistic habits die hard. Although from the Platonist point of view tenses are redundant, like the first and second persons, we do not concede that they altogether are. They may not convey information about the date of the event, E, or even the temporal perspective, given by Reichenbach's reference point, R, from which the event is to be viewed, but they show how both of these are connected with the date of utterance, S; and it is an essential condition of their being in time at all that some such connexion should obtain. Hence it is reasonable not to eliminate tenses even when for some purposes they are redundant. But if we continue to use them when they are supposed to be redundant we are easily ensnared in a sort of philosophical double-talk fertile of paradoxes and contradictions.

We can, if we insist upon it, speak timelessly of time. But we need to talk warily. It is an artifice, sometimes legitimate, but always unnatural, and our natural locutions are liable to trip us up. We can, if we like, make all our temporal, like all our spatial and personal, references explicit, so that our propositions are entirely independent of their context of utterance; and we can entertain such propositions, and wonder whether they are timelessly true or false, and which of them timelessly entail which other ones. But we must not then regard these timeless propositions as omnitemporally true (or false), and therefore as already true (or false) even before the event in question. *If* we adopt the tenseless mode of speech, then we cannot say that "*before* the stars saw one another plain" the death of Queen Anne *had* such and such causes and such and such effects [14], or that the Trojan war preced*ed* the Peloponnesian war, and equally of a proposition about any event that it *was* true 10,000 years *beforehand*, or that it *was* false 10,000 years *beforehand*, with the consequence that it had to happen as it did. This is not the only argument for fatalism. Nor, as we shall see, is it at all easy to talk so carefully about time that no fatalist argument can find any foothold. But fatalism is a fallacy to which those who talk timelessly about time are peculiarly prone.

[14]   McTaggart, op. cit., ch. 32.

§ 53
Dates and tenses

It is dangerous to mix the timeless and the tensed modes of speech, and it would seem wise either to follow Quine and keep ourselves exclusively to Iamblichus' intellectual time, or to construct a tense logic like Prior's on the basis of McTaggart's A series. But neither counsel can be followed through. I cannot, for the reasons I have given [1], discount time from my own stand-point altogether, and refuse to give any indication whatsoever of how I stand in relation to the topics talked of. But neither can I express myself adequately with sequences of past-, present- and future-tense operators, even if allowed to iterate them indefinitely often. For they suffer from an indefiniteness of refer-ence, which, although a natural economy of language, introduces a fatal vagueness into logic. I say 'I am coming', 'He will come', 'We have had supper', in contexts that make it adequately clear what range of dates is being referred to. And if on any occasion my hearer were uncertain whether I meant that he would come in the next half hour or the next six months, he could clarify the statement by asking me. In tense logic, however, as in ordinary logic, there is no provision for conversational elucidation. If I say 'He will come', I have to

[1] §§ 2, 7, 44, 52. See also Richard M. Gale, *The Language of Time* (London, 1968), esp. ch. IV.

be taken as meaning that he will come sometime. Prior, the pioneer of tense logic, accepts this consequence, and uses it to reinforce the parallel with modal logic. There is a deep analogy between the modal $\square$ or L, indicating necessity, and the logical term 'all', and between the modal $\lozenge$ or M, indicating possibility, and the logical term 'some'. In giving the future operator the modality of S4 and the past operator the modality of S4.3, we have assimilated P, more or less, to the $\lozenge$ or M of modal logic (see § 51, p. 267). But this interpretation requires us to construe P as 'it was the case *at some time*', where the italicized words *at some time* have the sense given by the existential quantifier in logic, $(\exists t)$. It follows that if we deny a statement beginning with the past operator P, we are asserting that it was *at no time* the case; and if we then prefix the same negated operator to a negated proposition, we are asserting that it was *at no time* the case that *not*, i.e. that it was *at all times* the case that.... Prior has besides the past operator P, corresponding to the modal $\lozenge$ or M, another past operator H, corresponding to the modal $\square$ or L. But it accords ill with ordinary usage. In ordinary usage the tense is much more part of the reference than of the predicate; it shows us to what event we are referring rather than gives us a further description of the subject. If I say Caesar crossed the Rubicon, I mean that Caesar crossed the Rubicon *at some time*, which I could specify further but do not need to; I do not mean by 'some time' merely *non nullo tempore* but *quodam tempore*. Similarly, if I say 'Peter was in lunch', I am speaking elliptically, and mean 'Peter was in lunch today', not that Peter at some time or other was in lunch. It is not a quantified statement with an existential operator ranging over time, but a singular statement whose temporal reference has not been specified very fully. The proof is in negation. If I were to deny that Caesar crossed the Rubicon, or that Peter was in lunch, I am not committed to claiming that Caesar never crossed the Rubicon (how could he have been in a position to cross it into Italy then?) or that Peter has never had lunch in college, but only that Caesar did not cross the Rubicon when you said he did, or that Peter was not in lunch today. In our ordinary way of speaking, we first find out what particular date is being referred to, and then if we deny the statement, deny it as referring to that date: we do not take it as having an indefinite temporal reference, and deny it by claiming that no such event ever occurred. Of course, if it were true that Caesar never had crossed the Rubicon, or Peter never had had lunch in college, it would be false that Caesar crossed the Rubicon in 49 B.C. or that Peter was in lunch today. But, in the ordinary way of speaking, these are contraries, not contradictories, and if I deny that Caesar crossed the Rubicon or that Peter was in lunch, it does not refute me, but is merely irrelevant, to say that Caesar did cross the Rubicon in 51 B.C. or that Peter was in lunch last Michaelmas term.

Tense operators should not be construed as *saying* that at some time (*non nullo tempore*) it was (or it will be) the case that ..., ..., but as *showing* that we are regarding an event at some time (*quodam tempore* – i.e. at some particular time we could specify) *as* a past event (or as a future event). It is, in Reichenbach's terms, an R-E calculus. We consider events from various reference points as past (if they are before the reference point), present (if they are more or less contemporary with it), or future (if they are after it), and consider what effect this variation of reference point has on the meaning and range of possible truth-values of the corresponding statements. It is the reference point not the date of utterance that is important in standard cases, because logic is concerned with what is invariant over different occasions of utterance. Thanks to Reichenbach's distinction we can separate the different temporal perspectives that should make a difference to a proposition from those we need to discount. Occasionally, as we shall see, the distinction is blurred. And it is these occasions that have given rise to the most puzzling problems about future contingents.

Let us assume we have established a dating system. Since we have a rational theory of clocks, we may assume dates given by real numbers. Let us further make it our convention that t_1 is before t_2, t_2 before t_3, etc. Our basic propositional unit will be about an event at t_2, and we can symbolize it

$$p_{t_2}.$$

If we regard p_{t_2} from an earlier date t_1, we shall have a future proposition which we shall symbolize [2]

$$(p_{t_2})_{t_1}.$$

Similarly we express past propositions by

$$(p_{t_2})_{t_3}$$

or, more generally, by

$$(p_t)_{t'}.$$

If we wish, we can carry out the same operation again to obtain further, more complicated, propositions, for example

$$[(p_{t_2})_{t_1}]_{t_3}.$$

We can construe these propositions in two ways. We may, first, construe them as the invariant core of tensed propositions:
$(p_{t_2})_{t_3}$ is what is common to

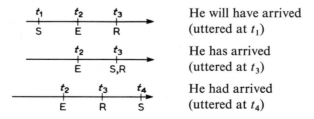

	He will have arrived (uttered at t_1)
	He has arrived (uttered at t_3)
	He had arrived (uttered at t_4)

[2]   I am indebted to the late Arthur Prior for suggesting this symbolism to me.

since the negation of any one of these, at the appropriate date, will contradict the others. If we regard them in this way, $(\ )_t$ can be taken as an analogue to the operators of modal logic. It operates on a proposition to produce another proposition, whose truth conditions are characteristically different from, although related to, those of the original proposition. In particular, we have the rule that if p_t is true, so is $(p_t)_{t'}$ for all t' such that $t' \geqslant t$.

Alternatively, we may consider that the different utterances 'It is going to rain tomorrow' made at different times are, on Reichenbach's analysis, expressions of different propositions, with the reference point being contemporaneous with the date of utterance in each case. The propositions are $(p_{t_8})_{t_1}$, $(p_{t_8})_{t_2}$, ... $(p_{t_8})_{t_8}$ where p_{t_8} is that rain falls at noon on Saturday 24 May 1969 and t_1, t_2, ... t_6 are dates on Friday 23 May and Saturday 24 May, and t_7 is 11.59 a.m. on Saturday 24 May. If we regard these propositions as being *about* the dates t_1, t_2, ... t_7, t_8, we can see how the fuller information available with the passage of time makes different propositional functions relevant, and different assignments of probability appropriate. We also see why so long as $t < t_8$ we may not be able to assign a definite truth-value but only a probability, whereas for all t such that $t \geqslant t_8$ we shall be able definitely to assign either truth or falsehood, since the specification of the t will include the fact either that rain did fall at noon on Saturday 24 May or (as the case may be) that it did not. The passage of time is constituted by an increase in information, and hence (except in certain cases which arise in quantum mechanics but do not concern us here) by a fuller and fuller specification of our topic of discourse. Once this is sufficiently full to include the answer to the question at issue, it is no longer an open question, but is definitely true or definitely false. This is the rationale of the rule that

for all t' such that $t' \geqslant t$, $(p_t)_{t'}$ has the same truth value as (p_t).

(Note that although in the simple case p_t must have the value 'true' or the value 'false', if p_t itself has only a probability – e.g. if p_t is itself of the form $(p_{t_2})_{t_1}$ – then $(p_t)_{t'}$ also will have only a probability.) This is what we often express by saying that the past is necessary, or, better, unalterable. Once t' is greater than t, the truth-value of $(p_t)_{t'}$ can no longer be altered, whereas while t' is less than t, $(p_t)_{t'}$ may either change or be changed in respect of its truth value. 'Necessary' is not a suitable term to characterize the past, because it is a characteristic thesis of all modal logics that $\sim \square p$ (or $\sim Lp$) is not equivalent to $\square \sim p$ (or $L \sim p$). Indeed, the addition of such a thesis to any modal logic collapses it into ordinary non-modal propositional calculus. But from our rule that $(p_t)_{t'}$ has the same truth value as p_t, if $t' \geqslant t$, it follows that since the Law of the Excluded Middle does hold for p_t – i.e. $\vdash p_t \vee \sim p_t$ – it also holds for $(p_t)_{t'}$ – i.e. $\vdash (p_t)_{t'} \vee (\sim p_t)_{t'}$. Hence it follows also $\vdash \sim (p_t)_{t'} \equiv (\sim p_t)_{t'}$. This is what we should expect – if it is not the case that it rained yesterday, then it is the case that it did not rain yesterday – but it means that the past-tense operator, 'it was the case that ...', is modally nugatory. The past tenses are straightforwardly indicative in mood, not mandatory or permissive. We therefore should not say that the past is necessary but rather that it is fixed, or, as of now, unalterable (although, of course, it might have been alterable before it happened).

The future is not fixed. If $t' < t$, $(p_t)_{t'}$ does not have to have the same truth value as p_t. Sometimes we may assign a probability to it; in other cases, either because of our ignorance, or because it is a matter for human decision rather than the natural tide of events, it is more appropriate to assign only the modalities possible, impossible, necessary or contingent. Rescher gives one intuitive way of mapping probabilities into modalities, but this gives S5 rather than S4. It is instructive to follow his method and see where it fails to serve our purposes. He assigns to $\Box p$ (or Lp) the value T if p has the probability 1, F if p has a probability in the interval $[0,1)$, and to $\Diamond p$ (or Mp) the value T if p has a probability in the interval $(0,1]$, F if p has a probability 0 [3]. From this it is easily seen that $\Diamond p \supset \Box \Diamond p$ (or $CMpLMp$) must always hold; and so the system is S5. What has gone wrong is that we have been considering the question whether p has a probability in $(0,1]$ or not from an abstract, non-temporal point of view, whereas on our account it will depend very much on one's temporal point of view what probability should be assigned to p, and hence whether it is 0 or falls in $(0,1]$. The modality that Rescher's rules yield is that of the omnitemporal physical possibility and physical necessity, not our feasible-at-a-given-time. As soon as we specify the given time, we see that we need to amend Rescher's rules. If $(p_{t_2})_{t_1}$ were to be assigned a particular probability, α, it would not follow that the same probability α should be assigned to it from the standpoint of every earlier time. Now that Montague and Sheila have seven daughters and have conceived another child, the probability that next year they will have eight daughters is about 0·48; but ten years ago when they married, the probability that ten years later the probability of them having eight daughters by the following year was 0·48 was itself less than 0·008. Similarly, now that I have married Jane, it is quite impossible that I should marry Charlotte too; but three years ago, when I was courting them both, it was perfectly possible that Charlotte might have been the girl of my choice, and if she had not given me the push it would have been possible then that marriage with her might be a possibility now. Neither the possibility then nor the impossibility now is an all-time possibility or impossibility, but only feasibility- or unfeasibility-at-a-given-time. In those cases where we can assess probabilities, we have a Markov chain. In the more general case, we have simply a sequence of possibilities and impossibilities, with no rule connecting different sorts of possibilities, but only that if ever it becomes impossible that p_t should be true, it must remain impossible at all subsequent times, and if ever

[3]   Nicholas Rescher, "A Probabilistic Approach to Modal Logic", *Acta Philosophica Fennica*, XVI (1963), pp. 215–25.

it becomes necessary that p_t should be true, it must remain so at all subsequent times.

We may call this view the "morning after the night before" analysis. For we take $(p_t)_{t'}$ as being itself a proposition about the *date t'*, saying that it was a t' which had followed (if $t' > t$) a t on which p had happened, or which was going to be followed by (if $t' < t$) a t on which p would happen. When we are dealing with future statements, e.g. that it is going to rain tomorrow, we find it natural to assign probabilities, but different ones at different times. When we hear on the wireless that a deep depression is on its way from Iceland, we assign a high probability, but a much lower one on seeing the red sky at sunset. Bright sun before seven the next morning sends the probability of rain up again, but it sinks as the wind veers round to the north-east at lunch time, and finally rises to unity as thereafter clouds blow up, and the drops begin to fall. These probabilities are being assigned to different propositions, each an instance of a propositional function 'a day on which the meteorological conditions are such-and-such will be followed by rain'. As time goes by, we have more and more meteorological information, and therefore can specify what sort of day it is more and more fully; and with each new specification the propositional function will range in a different universe of discourse and have a different probability assigned to it. And finally, the specification will include the fact that rain fell, and the probability will be unity, and will remain so for any subsequent specification that includes that fact [4].

In formalizing the calculus we return to the first view (p. 284–5), and seek analogues with modal logic. We need to have rules (1) connecting the truth-values of propositions about the same event, p_t, but from different reference points, $(\ )_{t'}$; and (2) connecting the truth value of some proposition $(p_t)_{t'}$ with the truth-value of a more complex one in which that proposition is itself regarded from another reference point – $[(p_t)_{t'}]_{t''}$. Our general principle is that it is all right to make either of these transformations if $t < t' < t''$ but not in general if $t > t' > t''$. We can combine them economically if we have a special rule about the present tense, namely the axiom

$$\vdash p_t \equiv (p_t)_t.$$

The effect of this is to allow iteration of tenses *via* the present, to give all propositional components, p_t, the status of propositions, and to give the present tense a peculiar, and ambiguous, status. This last point has been argued for, on quite independent grounds, by Prior [5]. Neither this nor the second point is in strict accord with Reichenbach's SRE analysis, but the argument for that analysis is its use in elucidating more complicated tenses, and, so far as the present tense is concerned, it seems somewhat artificial and

[4] See more fully J. R. Lucas, *The Concept of Probability* (Oxford, 1970), ch. VI, pp. 105–8.
[5] A. N. Prior, "On Spurious Egocentricity", *Philosophy*, XLII (1967), pp. 326–35; reprinted in A. N. Prior, *Papers on Time and Tense* (Oxford, 1968), ch. 2. See also A. N. Prior, *Past, Present and Future* (Oxford, 1967), ch. 1.

unnecessary. In the present tense the distinction between S, E and E is blurred, and it is reasonable to recognize this by the rule of equivalence between p_t and $(p_t)_t$. If this is granted, it carries with it the second corollary, that a propositional fragment, p_t, is itself a well-formed proposition in the present tense – a fact that makes formation rules much easier to state.

Formation rules:

I $p_t, q_t, r_t, \ldots$ are propositions, where t may be any real number that can be used as a date.

II   If A_t is a proposition, so is $\sim A_t$.

III   If A_t and $B_{t'}$ are propositions, so are $A_t \supset B_{t'}$, $A_t \lor B_{t'}$, and $A_t \,\&\, B_{t'}$.

IV   If A is a proposition, so is $(A)_{t'}$, where t' may be any real number that can be used as a date.

We have the axioms of propositional calculus, the rule of substitution, and the rule of *modus ponens*, together with the following additional axioms and rules:

A1 $\vdash p_t \equiv (p_t)_t$

A2 $\vdash (p_t)_{t'} \supset (p_t)_{t''}$ provided $t'' > t'$

A3 $\vdash (p_t)_{t'} \supset (p_t)_{t''}$ provided $t'' \geqslant t$

A4 $\vdash (p_t \supset q_{t'})_{t''} \supset [(p_t)_{t''} \supset (q_{t'})_{t''}]$

R5   If $\vdash A$ then $\vdash (A)_{t'}$.

We should note in A2 that the requirement is only that $t'' > t'$, and not that either should be greater than t. That is, A2 holds for future propositions as well as past ones. For, although the future is in general not fixed, sometimes it is, and if ever it is, it stays so until the time comes for the event described to happen. Therefore if $(p_{t_3})_{t_1}$ is true, $(p_{t_3})_{t_2}$ must be true also; which is as much as A2, $\vdash (p_t)_{t'} \supset (p_t)_{t''}$, says. For when $(p_t)_{t'}$ is not definitely true, the value F must be assigned to $(p_t)_{t'}$, and the axiom is satisfied vacuously [6]. By contrast, A3 is of importance only for past propositions, and is vacuous where future propositions are concerned. It says not only that past propositions are and always will be unalterable, but that they always have been ever since the events described. A2 and A3 together secure the modal vacuity of the past tense: that if it is not now true that it was so, it is now true that it was not so, and vice versa. For by propositional calculus

PC (Excluded Middle), $\vdash p_{t_2} \lor \sim p_{t_2}$

by A1, $\vdash p_{t_2} \supset (p_{t_2})_{t_2}$

by A2, $\vdash (p_{t_2})_{t_2} \supset (p_{t_2})_{t_3}$.

[6]   It is important here to distinguish the modal development of tense logic from the probabilistic. In the former there are only two truth-values, T and F; in the latter a continuous range from 0 to 1. If the probability that it will rain tomorrow is 8/9, then it is *not* true that it will rain tomorrow, and therefore to the proposition that it will rain tomorrow must be assigned the truth-value F. This does not mean, necessarily, that it is true that it will not rain tomorrow. See further below.

Hence, PC (Syll), $\vdash p_{t_2} \supset (p_{t_2})_{t_3}$.

Similarly, $\vdash \sim p_{t_2} \supset (\sim p_{t_2})_{t_3}$

$\therefore$ PC $\vdash (p_{t_2})_{t_3} \vee (\sim p_{t_2})_{t_3}$

$\therefore$ PC $\vdash \sim (p_{t_2})_{t_3} \supset (\sim p_t)_{t_3}$.

Conversely, by A3, $\vdash (\sim p_{t_2})_{t_3} \supset (\sim p_{t_2})_{t_2}$

by A1, $\vdash (\sim p_{t_2})_{t_2} \supset \sim p_{t_2}$.

Hence, PC (Syll), $\vdash (\sim p_{t_2})_{t_3} \supset \sim p_{t_2}$.

But, also by A1, $\vdash (p_{t_2})_{t_2} \supset p_{t_2}$

and, by A3, $\vdash (p_{t_2})_{t_3} \supset (p_{t_2})_{t_2}$.

Hence, PC (Syll), $\vdash (p_{t_2})_{t_3} \supset p_{t_2}$.

Hence, PC (Contraposition), $\vdash \sim p_{t_2} \supset \sim (p_{t_2})_{t_3}$.

Hence, PC (Syll), $\vdash (\sim p_{t_2})_{t_3} \supset \sim (p_{t_2})_{t_3}$.

If we did not have A3 as a separate axiom, we should not be able to argue from its now being true that it was not raining yesterday to its not now being true that it was raining yesterday. That is, there would be nothing wrong in saying both that it is now true that it was raining yesterday, and that it is now true that it was not raining yesterday. There is no similar difficulty with the future, because there A2 serves to ensure that we cannot have both that it is now true that it will rain tomorrow and that it is now true that it will not rain tomorrow, by bringing the latent contradiction to the surface at the later date, viz. tomorrow.

We may alternatively choose axioms to bring into prominence the analogue to the typical thesis of modal logic

$\vdash \Box p \supset p$

(or $\vdash Lp \supset p$)

instead of the special position of the present tense. We should then have

A3' $(p_t)_{t'} \supset p_t$

A2 $(p_t)_{t'} \supset (p_t)_{t''}$ provided $t'' > t'$

A1' $\vdash p_t \supset (p_t)_t$

A4 $\vdash (p_t \supset q_{t'})_{t''} \supset [(p_t)_{t''} \supset (q_{t'})_{t''}]$

R5   If $\vdash A$ then $\vdash (A)_{t'}$.

These two axiomatizations are equivalent. A1' follows immediately from A1. A3' can be proved thus:

if $t' \leqslant t$, by A2, $\vdash (p_t)_{t'} \supset (p_t)_t$

by A1, $\vdash (p_t)_t \supset p_t$.

Hence, PC (Syll), $\vdash (p_t)_{t'} \supset p_t$

if $t' > t$, by A3, $\vdash (p_t)_{t'} \supset (p_t)_t$.

Hence again, $\vdash (p_t)_{t'} \supset p_t$.

So in all cases, $\vdash (p_t)_{t'} \supset p_t$ A3'.

Conversely, A1 follows immediately from A1′ and A3′. A3 can be proved thus:

> by A3′, $\vdash (p_t)_{t'} \supset p_t$
> by A1′, $\vdash p_t \supset (p_t)_t$
> by A2, $\vdash (p_t)_t \supset (p_t)_{t''}$ where $t'' > t$.

Hence, PC (Syll), $\vdash (p_t)_{t'} \supset (p_t)_{t''}$ provided $t'' > t$.

It is easy to show that

> $\vdash \sim (p_{t_2})_{t_3} \equiv (\sim p_{t_2})_{t_3}$
> and $\vdash (\sim p_{t_2})_{t_1} \supset \sim (p_{t_2})_{t_1}$
> but not $\vdash \sim (p_{t_2})_{t_1} \supset (\sim p_{t_2})_{t_1}$.

The axiom A4 gives comparable rules for the tense operators and implication. It is the exact analogue of a thesis in modal logic. Essentially it entitles us to argue now from implications that we know either used to or are going to hold good. So too, R5, like the corresponding rule in modal logic, secures that the truths of logic held and will hold in past and future contexts. Given these and the rules for negation, we have a standard modal logic. So far as the past tense is concerned, it is trivial from the modal point of view. The future has the structure of S4. To show this we need to establish first the analogue of the thesis

$$\vdash \Box p \supset \Box \Box p$$
$$(\text{or} \vdash Lp \supset LLp).$$

This is $\vdash (p_{t_3})_{t_1} \supset [(p_{t_3})_{t_2}]_{t_1}$ for any t_2 such that $t_1 \leqslant t_2 \leqslant t_3$

> by A2, $\vdash (p_{t_3})_{t_1} \supset (p_{t_3})_{t_2}$ given $t_2 \geqslant t_1$
> by R5, $\vdash [(p_{t_3})_{t_1} \supset (p_{t_3})_{t_2}]_{t_1}$
> by A4, $\vdash [(p_{t_3})_{t_1}]_{t_1} \supset [(p_{t_3})_{t_2}]_{t_1}$
> by A1, $\vdash (p_{t_3})_{t_1} \supset [(p_{t_3})_{t_1}]_{t_1}$
> by PC (Syll), $\vdash (p_{t_3})_{t_1} \supset [(p_{t_3})_{t_2}]_{t_1}$.

We should notice that the result is slightly stronger, in that t_2 is not required to be $\leqslant t_3$. That is right. If it will be the case that p the day after tomorrow, it follows both that it will be the case tomorrow that it will be the case that p the day after, and that it will be the case in three days time that it was the case that p the day before. The future implies not only the future future but the future perfect. With the past tense, however, whereas it is trivially easy to argue from the simple aorist to the pluperfect, there is no argument from the simple aorist to a past-future tense. If it rained the day before yesterday then it was the case yesterday that it had rained the day before, but it does not follow that it was the case three days ago that it was going to rain the following day. A large part of the labour of tense logic is to find a formulation in which that unwelcome inference does not hold.

That the future operator does not have the modal logic of S5 can best be seen by considering its interpretation. The characteristic thesis of S5 is

$$\vdash \Diamond p \supset \Box \Diamond p$$
$$(\text{or, } \vdash Mp \supset LMp);$$

in our interpretation, if it is feasible that p then it is fixed that it is feasible that p.

$$\sim (\sim p_{t_3})_{t_2} \supset [\sim (\sim p_{t_3})_{t_2}]_{t_1}.$$

But this is not so. It may be feasible now for me to marry Jane. But it will not go on being feasible, if I invite Jean to come with me to the dance; nor was it a foregone conclusion that the opportunity would be open to me now – if I had taken Jean out last year, which I perfectly well could have done, it would not have been feasible for me now to win Jane's hand in marriage. It was not fixed that it would be feasible, any more than it is now fixed that it will continue to be feasible. All that is fixed by its being feasible now is that it will be the case hereafter that it *was* feasible, i.e.

$$\sim (\sim p_{t_3})_{t_1} \supset [\sim (\sim p_{t_3})_{t_1}]_{t_2}$$

and this does not exemplify S5, as the operators are of different tenses. Similar examples rule out other modal logics between S4 and S5, so far as the future operator is concerned.

In this section we have sketched what might be called an R-E calculus, concerned with the invariant features of tensed discourse. The most characteristic features, however, are those that depend on the occasion of utterance, S. It is difficult to capture this feature, just because it depends on context, and logic seeks to be context-invariant. Something can be achieved by following Prior's suggestion that S should be regarded simply as the first of a number of reference points, R_1, R_2, …. Certainly sometimes the distinction between S and R is blurred, just as that between R and E is. In particular, this, combined with the difficulty of handling the logical relations between utterances made on different occasions, has helped generate the many fallacies about future contingents and fatalism.

§ 54
Future contingents and fatalism

There are many arguments for fatalism. They all depend on the interplay of
utterances made on different occasions in different tenses, and how some of
these are naturally described as being true. They are all fallacious. But although
they are all rationally invalid, we have great difficulty in agreeing exactly where
the fallacious step is made. For a number of different ambiguities overlap,
each reinforcing the confusion the others generate, and each made more difficult
to detect by reason of the others. It is tempting, then, to remove the ambiguities
by *fiat*, and stipulate that certain key words, such as 'true', be used only in
specified ways. But a price has to be paid. The ambiguities are not accidental.
They reflect natural needs. If we straitjacket the words, although we block the
fallacious inferences to fatalism, we also block many inferences that seem
natural and legitimate. Rather than stipulate how words shall be used, we
should attempt to see more clearly why they are used as they are, and how
differing pressures in different directions may be reconciled.

Three crucial ambiguities confuse us: the correct analysis of the future
simple tense; the meaning of the word 'true'; and our rules for rendering *oratio
recta* into *oratio obliqua* and back again. Most European languages, including
English and Greek, have only one future form, in contrast to their two past
forms, the aorist and the perfect: we do not easily distinguish between the cases

+——————→ and ├————┤——→ and part of the problem of future
S, R   E   S   R, E

contingents has been generated by the unconscious transition from the latter future simple tense, which has no sense of the future being already fixed, to the former "perfect future" ("posterior present" in Reichenbach's terminology) where the future is "already present in its causes", at least as far as language goes. When I am making a prediction, I use the future in the latter sense. For a prediction is not simply a conjecture or a guess. The mere fact that I am making an assertion in the indicative mood carries with it, in the absence of any further indication to the contrary, a presumption that I am doing so with reason. Unless I make it clear that I am only speculating idly when I say Peter will be at the party, you are entitled to assume from the mere fact of my saying it that I have some reason for believing that Peter is going to be at the party. And I must have my reasons at the time of uttering the sentence, not after the party has actually taken place. Hence, to predict is to use the (S, R-E) future tense, which I shall express by the auxiliary verb 'is going to'.

In other cases, and in particular in other moods, the future should be analysed as being of the (S-R, E) form. If I bet you that Eclipse will win the Derby, I am not warranting that he is going to: I am staking only my money, not my reputation, on the event. If he wins, my guess turns out to be correct, but its correctness depends entirely on whether he wins, not at all on the soundness of my reasons for backing him. If, on being asked why I had backed him, I gave an entirely frivolous reason, or no reason at all, there would be nothing wrong with my bet or my having made it; whereas not to be able to give faceable reasons does show a prediction to have been misleading, irresponsible and unwarranted, even if not falsified in the event [1]. When I ask questions, and wonder whether I shall catch the train, or make plans, and decide what to do if I should miss the train, I have no stake in the present. Although I am speaking at S, my mind is on E, and therefore R is contemporaneous with E, not S. Either I shall catch the train or I shall not. If I catch the train (here the English use of the present tense is revealing), I can go to the library and look up some references before going back to college for dinner; if I miss the train, I can go to the municipal art gallery until the next train comes.

If we accept the R-E calculus of the previous section, the difference between (S, R-E) and (S-R, E) becomes clear and important. The Law of the Excluded Middle applies straightforwardly to the latter case, but not the former. When I am making contingency plans, I can argue "Either I shall catch the train or I shall not", for, although I am speaking now, I am referring to a future date, the time of the train's departure, when either I am on it, or I am not. Although the relation between the S and the E requires me to use the future tense, the relation between the R and the E is one of contemporaneity, and therefore makes the

[1]   See further S. E. Toulmin, *The Uses of Argument* (Cambridge, 1958), pp. 57–62.

logic straightforward. I cannot predict with equal confidence that I am going to catch the train or that I am going to miss it. For it may still depend on how fast I run or how slow the traffic is. We cannot say now, at t_1, $\vdash (p_{t_2})_{t_1} \vee (\sim p_{t_2})_{t_1}$, but only the weaker $\vdash (p_{t_2})_{t_1} \vee \sim (p_{t_2})_{t_1}$. And as we have maintained in § 53, the latter does not entail the former, because $\sim (p_{t_1})_{t'}$ does not in general imply $(\sim p_t)_{t'}$ when $t' < t$. I can say that either I am definitely going to catch my train (e.g. if I am at the station ten minutes before it is due to depart) or I am *not definitely* going to catch my train, but not that either I am definitely going to catch my train or I am *definitely* going *not* to catch it.

Aristotle alludes to this argument in his famous discussion of future contingents in *De Interpretatione*:

εἶναι μὲν ἢ μὴ εἶναι ἅπαν ἀνάγκη, καὶ ἔσεσθαί γε ἢ μή. οὐ μέντοι διελόντα γε εἰπεῖν θάτερον ἀναγκαῖον. λέγω δὲ οἷον ἀνάγκη μὲν ἔσεσθαι ναυμαχίαν αὔριον ἢ μὴ ἔσεσθαι, οὐ μέντοι γενέσθαι αὔριον ναυμαχίαν ἀναγκαῖον οὐδὲ μὴ γενέσθαι. γενέσθαι μέντοι ἢ μὴ γενέσθαι ἀναγκαῖον.

Everything necessarily is or is not, and will be or will not be; but one cannot divide and say that one or the other is necessary. I mean, for example: it is necessary for there to be or not to be a sea battle tomorrow; but it is not necessary for a sea battle to take place tomorrow, nor for one not to take place – though it is necessary for one to take place or not to take place. [2]

Aristotle's discussion is vitiated by a confusion between the different senses of ἀναγκαῖον involved. If it is logical necessity, it is clearly illegitimate to divide it and argue that either it is logically necessary for a sea battle to take place or it is logically necessary for one not to take place; but although this fallacy could be committed, it is not the one we are most tempted to commit. The modal operator we are tempted to divide is not a logical 'must', but a temporal 'shall' or 'is going to', and to argue from the true fact that tomorrow there is going to be either a sea battle or not, to the false conclusion that either now there is going to be a sea battle or now there is not. And this argument is made easy because of the ambiguity, natural and inevitable though it is, between the (S, R-E) and the (S-R, E) future tenses. Speaking at t_1, I can properly say both

$$(p_{t_2} \vee \sim p_{t_2})_{t_1} \quad \text{and} \quad (p_{t_2} \vee \sim p_{t_2})_{t_2}.$$

The latter, but not the former, can divide its modality, i.e. the latter entails $(p_{t_2})_{t_2} \vee (\sim p_{t_2})_{t_2}$. It is then easy to suppose that the former must likewise entail $(p_{t_2})_{t_1} \vee (\sim p_{t_2})_{t_1}$ and the fallacy follows.

Aristotle's main argument, however, is given earlier and concerns the word 'true' – ἀληθὲς εἰπεῖν, true to say. The word 'true' is connected both with actual utterances asserted on particular occasions and with the invariant content of what is or might be asserted, independently of any actual assertion

[2]   *De Interpretatione*, tr. J. L. Ackrill (Oxford, 1963), ch. 9, p. 53, 19a 28–32.

in an actual context. Etymologically it is connected with 'trust' and 'troth'. If I say something is true, I am promising that it is trustworthy. It carries with it, only much more strongly, the same warranty of my having good reasons as my simple assertion does. 'It is true that ...' has the opposite effect from 'It is probable that ...'. The latter reduces or withdraws the warranty that what is said can be relied upon, the former gives an extra guarantee. This extra guarantee is so great that it becomes unconditional. If I say that it is true that I shall catch the train, as I arrive at the station with ten minutes to spare, and then the ticket clerks go on strike or the train is derailed as it approaches the platform, I was wrong, even though I had excellent reasons for giving the guarantee. With 'probable', whether I was right or wrong to make a probabil- istic statement depends entirely on the reasons I had for making it, and not at all on how things turned out in the event. With 'true' it is the other way about; although I ought to have reasons, and if they are sufficiently good I may not be blamed if I am proved wrong by events, yet when the crunch comes my state- ment cannot stand if things do not turn out as predicted. The word 'true', which started as an operator at the time of utterance, S, has come to be an evaluation which stands or falls by what actually happens in the event, E. Instead of applying to a specific *utterance* made by a particular person at a particular time – e.g. 'It will rain tomorrow' said by you yesterday – it applies to what is common to a whole set of suitably different utterances at different times; the *proposition* that could have been expressed by anyone yesterday uttering the sentence 'It will rain tomorrow' or today 'It is raining today' or tomorrow 'It rained yesterday' or at any time after today by 'it rained on 24 May 1969', or .... According as the word 'true' is used as an operator applied by a particular person to an actual utterance at a particular time or as a truth-value ascribed to a proposition which is independent of person, occasion of utterance and exact formulation, it assimilates the reference point, R, to S or to E, and thus reinforces the ambiguity between (S,R-E) and (S-R,E) already noted.

In most languages we use the phrase 'it is true that' with indirect speech, *oratio obliqua*, in exactly the same way as 'it is said that'. But when we use indirect speech to report actual speech, there is an automatic shift from the original speaker's date of utterance, S, to a new reference point, R_1, which *we* are referring to. If you said yesterday 'It will rain tomorrow', I shall report you today 'You said that it would rain today'. The 'would' refers to the rain from the standpoint of yesterday when you uttered the words 'It will rain tomorrow'. Your S has become my R, for what I am talking about is the fact of your having uttered words to a particular effect. Linguistic usage requires a similar shift with 'It is true that ...', but often, especially in sophisticated uses, the word 'true' is concerned very little with the actual utterance, and linguistic pressure to introduce an extra reference point is misleading. If I say, 'It was true that it would rain today', the word 'true' may be not an operator governing the

utterance 'It will rain tomorrow', which was or might have been uttered yesterday, but a truth-value assigned to the proposition that could have been
expressed yesterday by the sentence 'It will rain tomorrow', or today by the
sentence 'It is raining today', or tomorrow by the sentence 'It rained yesterday',
or at any time after today by 'It rained on 24 May 1969', or ..., or .... In that
case, the truth of the proposition will stand or fall by whether it rains today, and
nothing else; and therefore there is neither any need to date its truth to yesterday, nor any possibility of doing so without creating confusion. Propositions
have been devised to be independent of time, and therefore we should not say
that a proposition was true or will be true, but only that it is (tenselessly) true.
And rather than use *oratio obliqua* with its insinuation of a redundant and
misleading reference point, we do better to use the apparatus of quotation
marks and eternal sentences after the manner of modern logicians.

The double ambiguity, between the two future tenses, and the two uses of
the word 'true', lie behind Aristotle's main argument for fatalism:

ἔτι εἰ ἔστι λευκὸν νῦν, ἀληθὲς ἦν εἰπεῖν πρότερον ὅτι ἔσται λευκόν, ὥστε
ἀεὶ ἀληθὲς ἦν εἰπεῖν ὁτιοῦν τῶν γενομένων ὅτι ἔσται. εἰ δ᾽ ἀεὶ ἀληθὲς ἦν
εἰπεῖν ὅτι ἔστιν ἢ ἔσται, οὐχ οἷόν τε τοῦτο μὴ εἶναι οὐδὲ μὴ ἔσεσθαι ....

If something is white now, it was true to say beforehand 'it will be white';
and thus of a past event it was always true to say 'it will be'. If it was always
true to say 'it is' or 'it will be', it could not be that it neither was, nor was
going to be, the case. It is impossible that something should not happen
which is such that it could not be that it would not happen. If it is impossible
for it not to happen, it is necessary that it should happen. With all future
events, therefore, it is necessary that they should happen. [3]

The words ἀληθὲς ἦν εἰπεῖν ("it was true to say") reinforce the equivocation
between what was uttered and the act of uttering it. If what was uttered was
the sentence ἔσται λευκόν ("it will be white") in the future simple sense (S-R, E),
then it expresses the same proposition as ἔσται λευκὸν νῦν ("it is white now"),
said at the same time as the event, or ἦν λευκόν ("it was white"), said afterwards.
And if the latter two express a true proposition, then so does the former. But
the utterance is true only by virtue of the proposition it expresses, not on
account of any merit that inheres in itself. Although it was uttered at a particular time, truth was not one of its properties then. Truth, in this sense, attaches
to the proposition, not the utterance, and whether the utterance had succeeded
in expressing a true proposition or not was not finally settled by reason of the
utterance's having been irrevocably uttered. If, however, we turn our attention
from what was uttered to the uttering of it, as the ἦν (was) and the εἰπεῖν (to
say) encourage us to do, then we construe it as having been an actual prediction
in the (S-R, E) future tense, and assess it with reference to the date on which it

[3]   *De Interpretatione*, ch. 9, 18b 9–15.

was uttered. If you, on 23 May 1969, actually uttered the words 'It will be white' (say, of a wedding), you imply that you have reason to believe it (you have seen the dress, or the bride's mother told you), and therefore the future is indeed present in its causes, and the (S-R, E) analysis is correct. And this justifies our rule in *oratio obliqua* that requires me to report you, on 24 May, as having said that it would be white, and if I furthermore endorse this as being true there is very strong linguistic pressure to refer the truth to the date of your saying it. And thus, it seems that it was then already the case that it was going to be true, and so had lost all chance of not being true, and simply had to be true.

To put it diagrammatically: if today, Saturday 24 May 1969, 'it is white'

$$\xrightarrow[\text{S,R,E}]{\text{24 May}}$$ is true, then by the ordinary rules for different utterances in

oratio recta at different times, the utterance 'it will be white', if it had been made yesterday, would have expressed the same proposition, which *is* (tenselessly [4]) true. That is, if

$$\xrightarrow[\text{S,R,E}]{\text{24 May}} \quad is \text{ true, so } is \quad \xrightarrow[\text{S} \qquad \text{R,E}]{\text{23 May} \quad \text{24 May}} \, .$$

But if the potential utterance 'It will be white' *is* true, so would have been the actual utterance, had it been made, and this we report, on 24 May, as

$$\xrightarrow[\text{R}_1 \qquad \text{S,R}_2\text{,E}]{\text{23 May} \quad \text{24 May}} \quad \text{or on 26 May} \quad \xrightarrow[\text{R}_1 \qquad \text{R}_2\text{,E} \qquad \text{S}]{\text{23 May} \quad \text{24 May} \quad \text{26 May}}$$

in which R_1 refers to the date of the utterance and R_2 to the date that the utterance itself referred to. In either case there is a natural tendency to overlook R_2, since E obviously is the event in question, and to construe these reports simply as

$$\xrightarrow[\text{R} \qquad \text{S,E}]{\text{23 May} \quad \text{24 May}} \quad \text{and} \quad \xrightarrow[\text{R} \qquad \text{E} \qquad \text{S}]{\text{23 May} \quad \text{24 May} \quad \text{26 May}} \, .$$

But these in turn are ambiguous, and could equally well be oversimplified versions of

$$\xrightarrow[\text{R}_1\text{,R}_2 \qquad \text{S,E}]{\text{23 May} \quad \text{24 May}} \quad \text{and} \quad \xrightarrow[\text{R}_1\text{,R}_2 \qquad \text{E} \qquad \text{S}]{\text{23 May} \quad \text{24 May} \quad \text{26 May}}$$

which are the rendering into *oratio obliqua* of the utterance $\xrightarrow[\text{S,R} \qquad \text{E}]{\text{23 May} \quad \text{24 May}}$

'it is going to be white'. The pull of the words ἦν (was) and εἰπεῖν (to say) is towards this interpretation, since they make us think of the dated utterance,

[4]   I borrow this useful convention from Richard M. Gale, *The Language of Time* (London, 1968), p. 18, n. 2.

which, if it actually occurred, was a prediction and therefore of the (S, R-E)

form. Hence we have argued from an innocuous $\xrightarrow[S,R,E]{24\,May}$ to an ambigu-

ous $\xrightarrow[\;\;R\;\;\;\;\;\;S,E\;\;]{23\,May\;\;\;\;24\,May}$ which we verbalize as 'it was going to be', today's

equivalent of yesterday's $\xrightarrow[\;\;S,R\;\;\;\;\;\;\;\;E\;\;]{23\,May\;\;\;\;\;24\,May}$. The ambiguity of ordinary

language enables us to confound the two future tenses, while that of the word 'true', which in turn is made greater by Aristotle's phrase, forces us to refer both to the date of the event, which tells us whether the proposition is in fact true, and to the date of the hypothetical assertion, which, being before the event, and yet true, suggests that it must have been a foregone conclusion.

We can avoid the fallacy if we take care to distinguish the rules governing the variable dates of utterance, S, in *oratio recta* from the reference point, R, which is invariant in *oratio recta*, and should be so in oblique contexts too, and also to distinguish the different uses of the word 'true'. Then, granted the R-E calculus of the previous section, we can escape the fatalist conclusion. Aristotle had dimly discerned this. He denied that the Principle of Bivalence (that every proposition is either true or else false) applied to every proposition about the future. And on our analysis, too, if we consider propositions expressed by 'is going to' (S, R-E) sentences, no exception need be taken. There is no reason why it should be either true, as of now, that something is going to be the case or true, as of now, that it is not going to be the case. Neither need be true *as of now*. What is not tolerable, however, is that I should now be saying, with reference to the date of the putative event, that it *then* need neither be the case nor not be the case. If I am considering a point of time, R, then, whatever the actual date I am speaking at, I must allow that either E happens at R or it does not. A date that could be neither a date at which E happens nor a date at which E does not happen is an altogether inconceivable date. The Principle of Bivalence must apply to propositions in which the date referred to is con-temporaneous with the state of affairs alleged to obtain at that date. It is only when the date referred to is not contemporaneous with the putative state of affairs (and then only if it is before) that we enter a *caveat* and say that the negation of 'is going to' is not 'is going not to' but 'is not going to'. Aristotle thought that the Law of the Excluded Middle did apply to future propositions, but that the Principle of Bivalence did not: that is, that 'There is going to be either a sea battle or no sea battle tomorrow' is a truth of logic, whereas 'Either it is true that there is going to be a sea battle tomorrow or it is false that there is going to be a sea battle tomorrow' is false.

The same solution has been put forward recently by Cahn, who distinguishes what he calls the analytic law of the excluded middle (corresponding to what I have called simply the Law of the Excluded Middle) from what he calls the synthetic law of the excluded middle (corresponding to what I have called the Principle of Bivalence), and who admits the former as a law of logic, while denying the latter as being merely synthetic and actually untrue [5]. We can agree, if we construe (S, R-E) sentences as expressing R-E propositions which may have any probability value in the interval [0, 1]. We do not assign to the proposition that the next toss of a coin will come down heads either the value 1 (true) or the value 0 (false), although we do assign the value 1 (true) to the proposition that the next toss of a coin will either come down heads or not. So too we need not assign to the proposition that it is going to rain tomorrow either the value true or the value false, but something in between. But this is because we are choosing to work with a many-valued logic. We can, equally well, insist on the Principle of Bivalence, and regard R-E propositions as complex ones, in which the R acts as a modal operator on the E. In that case the Principle of Bivalence applies, but the rules for negation are complicated. From its being false that there is going to be a sea battle tomorrow, it does not follow that there is going to be no sea battle tomorrow. Unless we have passed the moment of decision, and the die is already cast, it may be the case that neither 'There is going to be a sea battle tomorrow' nor 'There is going to be no sea battle tomorrow' is true.

Aristotelian solutions have not found favour with philosophers, because they contravene the principle that time is not itself "efficacious" [6]. A mere difference of temporal perspective makes all the difference to whether a proposition is true or false, or whether a modal operator commutes with negation; whereas it is generally held that a difference of time should make no difference (see §§ 12, 14). But the principle of date-indifference is a principle of natural science, not of logic. Scientific laws should be omnitemporal, and be no respecters of persons or dates. But tense logic is by its nature concerned with times and differences of date. There is no reason why it should be date-indifferent; indeed, if it were, it would not be *tense* logic at all. For particular scientific purposes we construe time as being tenuous and featureless, and disregard those profound characteristics of the passage of time that we are all aware of in our capacity as agents (see § 15). It is reasonable to do this in the prosecution of science, but we should regard it as a limitation on the scope of science, and not as a scientific discovery about time. Temporal perspectives can be discounted for some purposes, but not for all, for it is part of being human to be an agent, and all actions are undertaken *sub specie temporis*.

[5] Steven M. Cahn, *Fate, Logic and Time* (New Haven and London, 1967).
[6] Richard Taylor, "Fatalism", in *Metaphysics* (Englewood Cliffs, 1963); reprinted in *Philosophical Review*, LXXI (1963); and in Richard M. Gale (ed.), *The Philosophy of Time* (London, 1968), p. 224.

§ 55
Eternity

Eternity is much misunderstood. The ordinary man is frightened when he contemplates everlasting aeons stretching on world without end down the corridors of infinite time; and the philosopher, and too often the theologian also, takes refuge in a Platonic doctrine of timelessness, which only bypasses problems and does not solve them. Platonism apart, there is an argument from the changelessness of God's substance that has often misled philosophers into extruding God from time altogether. The spatial analogy and a confusion between three different senses of the word 'present' confused St Augustine and Boethius, and not only them. And in all our thought about God, our urge to think of Him as unfettered from the limitations of our human condition, among which temporal limitations loom large, has often led us to overstep not the limits of human life, but of intelligibility.

Eternity is not timelessness. For eternity is an attribute of God, and God is a person, a conscious personal being, and time is an inevitable concomitant of consciousness (see § 2, pp. 8–9). To say that God is outside time, as many theologians do, is to deny, in effect, that God is a person [1]. The Absolute, τὸ ὄν, the Form of the Good, or even, perhaps, the Ground of our Being, may

[1] See further and more fully W. C. Kneale, "Time and Eternity in Theology", *Proceedings of the Aristotelian Society* (1960–1), pp. 87–108.

be outside time, and timeless in a full-blooded Platonic sense, but they are not persons: they neither see what we are, nor hearken unto our prayers, nor care what we do, let alone ever intervene in the course of the world's events. If we think of God as a living person, who acts in the world, or even who is merely conscious, we must seem to be ready to apply temporal expressions to Him, because the applicability of temporal predicates of some sort or other is a necessary condition of activity, even the inactive activity of consciousness. A timeless Deity may be the Truth; it may possibly provide us with the Way, or at least with a Goal; but it cannot ever be the Life. It remains necessarily only τὸ θεῖον not ὁ Θέος. Even Plato, in the *Timaeus*, has to distinguish the Demiurge from the Forms. To be alive, to be a person, to be conscious, to be active, one must be, in some sense, in time. Verbs can conjugate in persons only if they can conjugate in tenses too.

Nor is eternity changelessness, as theologians have understood the term. We need to distinguish two senses, an ordinary, relative one, and a metaphysical, absolute one, used by philosophers when talking about substances, or fundamental things. We often, when speaking of ordinary people or institutions, want to ascribe some sort of changelessness to them, and call on the breezes to waft wide "Our glorious *semper eadem*, the banner of our pride". The old school is still the same, we find, when we go back there after many years. We meet an old friend, and exclaim that he has not changed one bit. But we do not mean by this that he has been in a state of suspended animation since we last saw him. We mean that his personality has not changed, that he has lost none of his endearing traits of character. But he is quite likely to have changed his income bracket, and may have bought a new suit or a new car; and it would be impossible for him to reveal the old flashes unless he was moving and talking and saying new things, and in that sense changing.

The word 'changeless' in this sense applies to persons and institutions in much the same way as 'permanent' does to things and matter (see §§ 13, 19, 27, 28, 30, exp. pp. 74–5, 105, 126, 129–30, 142). We take into account only some features, not all features. The friend's character is unchanged, although his vocal chords are, as always, in continual motion. The atoms of the materialists are unchanged, unaltered, uncreated, indestructible; yet matter can move – indeed, must be allowed to, if it is to perform its metaphysical function. Even in philosophy, the argument for the permanence of substance establishes the need for only a relative permanence (§ 13, p. 75; § 27, p. 126). An absolute changelessness is no use to the metaphysician. His fundamental substance, whatever it is, must account for the fact of change. Parmenides, like Heracleitus, is out of court. The theist cannot have his fundamental substance, God, absolutely changeless, any more than the materialist can make his atoms static and incapable of motion. The permanence of matter and the changelessness of persons neither needs to be nor can be absolutely absolute. God may not wax old as doth a garment, or perish or be changed as a suit of clothes is changed and cast off [2]. But this is not to say that God cannot alter in any respect, and cannot be moved and cannot act. We may sing 'O Thou, that changest not, abide with me' but equally we ask Him to do things, which would be pointless if He could not act or alter in any respect whatever. Theologians, however, are tempted to wish an absolute changelessness on God. God, they say, is the same yesterday, today and for ever, not only in all essential respects, but in all respects whatsoever. But such a changelessness is open to the same objection as timelessness, namely that it is incompatible with personality. Personality is dynamic, not absolutely static. God may stand as firm as a rock, but is not as unfeeling as a rock. We ascribe all the permanence and stability that we want to ascribe to God by means of the first sense of changeless. It is only an improper use of the theological superlative and St Augustine's failure to rethink the Neoplatonist doctrines he had earlier espoused [3] that has led us to speak of God in an incoherent way, which makes Him out to be dead and finished instead of live and ever new. If matter, for all its permanence, must be allowed motion none the less, God, changeless and the same yesterday, today and for ever, uncreated, unaltered and indestructible, must be allowed to change nonetheless, if He is to be God at all. But still external change, imposed as it were from outside, seems to derogate from the transcendence and omnipotence of God. It is a point St Augustine returns to again and again: " *Si enim recte discernuntur aeternitas et tempus, quod tempus sine aliqua mobili mutabilitate non est, in aeternitate autem nulla mutatio est, quis non videat quod tempora non fuissent nisi creatura fieret quae aliquid aliqua motione mutaret. ...*" [4] "*In aeternitate stabilitas est, in tempore autem varietas: in aeternitate omnia stant, in temporalia accedunt, alia succedunt.*" [5] And Bede glosses the χρόνος οὐκέτι ἔσται [6] of the Apocalypse as meaning that at the last trump *mutabilis secularium temporum varietas ... cessabit* [7].

The Eastern Church took a much less static view. Gregory of Nyssa had a strong sense of the life of the soul being a perpetual progress, an *epectasis*, and was always quoting Philippians III, 13, to show that we should always be reaching out to goals that lie ahead [8], although the endless merry-go-round of mere physical change will not get us anywhere [9]. The definition of St

[2]　Psalm 102, 26–7.
[3]　St Augustine never repudiated Neoplatonism with the same vigour as he rejected Manicheeism and paganism (*Confessions*, VII, ii–v; see also *Contra Academicos*, XIII, 18). His account of eternity is reminiscent of that of Plotinus (see e.g., *Enneads*, III, 7, 3, ll. 36–8, and ch. 11 and 12) and more appropriate to an impersonal Absolute than a living God.
[4]　*De Civitate Dei*, bk XI, ch. vi.
[5]　*Sermones*, CXVII, 10 (vii).
[6]　Revelation, X, 7: "There shall be no more delay" (New English Bible).
[7]　Bede, *Explanatio Apocalypsis*, in J. A. Giles (ed.), *Bedae Opera Omnia* (London, 1844), vol. XII, p. 383.
[8]　See, e.g., *Life of Moses*, in J. P. Migne, *Patrologia Graeca*, vol. XLIV, 400D–401B (ed. J. Daniélou, II, 224–7); or *Commentary on the Song of Songs P.G.*, XLIV, 940D–491A.
[9]　See, e.g., *Life of Moses*, *P.G.*, XLIV, 344A (ed. J. Daniélou, II 60–1); or *On Ecclesiastes*, *P.G.*, XLIV, 624D–625A and 648D.

Maximus Confessor, although reminiscent of St Augustine, preserves the temporal character of eternity. He defined eternity not as being timeless, but as time in the absence of change:

Αἰὼν γάρ ἐστιν ὁ χρόνος, ὅταν στῇ τῆς κινήσεως, καὶ χρόνος ἐστὶν ὁ αἰὼν, ὅταν μετρῆται κινήσει φερόμενος, ὡς εἶναι τὸν μὲν αἰῶνα, ἵνα ὡς ἐν ὅρῳ περιλαβὼν εἴπω, χρόνον ἐστερημένον κινήσεως, τὸν δὲ χρόνον αἰῶνα κινήσει μετρούμενον.

For time is eternity, whenever it stands apart from change, and eternity is time whenever it is measured by being the vehicle of change. To put it in a nutshell, eternity equals time without change, and time equals eternity being measured by change. [10]

Eternity is that tranquil state envisaged by Barrow, when all the stars were still and only the perfect mind of God perceived the passage of time. Or as we should put it, eternity is time conceived as the concomitant of consciousness, and not as the dimension of change. It is an attractive doctrine, but too solipsistic to be really Christian. It makes eternity intelligible at the dawn of creation as the *distentio animi* of a self-sufficient creator, but does not allow for there being other persons, each with a mind of his own, who can act on his own and thereby initiate changes that are external, and might be unwelcome, to other minds. The Neoplatonist doctrine of the Impassibility of God is incompatible with the Christian doctrine of the Passion of Christ: we cannot secure that eternity shall be free of unwelcome change by stipulating that it shall be without change altogether, or we shall no longer be able to engage in activities and be fully personal. Therefore eternity is not simply time without change. Nevertheless, St Maximus' insight is valuable. It is characteristic of unredeemed time that its passage is accompanied by unwelcome change imposed from outside, and it is a mark of eternity that changes should be of the nature of activities, either initiated by each person or at least accepted by him as being in conformity with his will. In a non-solipsistic universe it is feasible only if there can be achieved a complete harmony of wills, and, on the Christian understanding, such a harmony cannot be quickly, easily or painlessly secured. But the concept of eternity is intelligible, at least for God, because for Him changes are not mere external movements that do not really matter, but are emotions that impinge on, and actions that spring from, the heart, from the very core of His Being.

Another reason why we often misunderstand eternity is that, as with time, we are misled by the spatial analogy. We think that all space must be present to God, and argue that all time must be present too, from which, confusing instants with intervals (see § 3), we conclude that God's time must be essentially static. It is clearest with Boethius. In his *De Trinitate* he takes us straight

[10]   St Maximus Confessor, *Ambiguorum Liber*, VI, 31 in J. P. Migne, *P.G.*, XCI, 1164B 14–C3.

through the categories one by one, first place and then *eodem praedicatur modo*, '*Quando*' [11]. As regards place, Boethius says, quite rightly, "*De deo vero non ita, nam quod ubique est ita dici videtur non quod in omni sit loco (omnino enim in loco esse non potest) sed quod omnis ei locus adsit ad eum capiendum, cum ipse non suscipiatur in loco.*" God is not to be located here, there, or everywhere, but everywhere is present to God, held by him, known to him. This is the lament of Psalm 139, and it was this that Newton was trying to express in his doctrine that space was the Sensorium of God. It is a natural and intelligible doctrine because if we take it upon ourselves presumptuously to think ourselves in God's position and to take a God's-eye view of the universe, it seems evident that everything must be present to omniscience, and that nothing can be dim or distant or remote. The Greeks, before Plato and Parmenides, had thought of God as changeless (rather than timeless), not moving because He, omnipotent, did not need to move in order to accomplish His will:

$$αἰεὶ \ δ' \ ἐν \ ταὐτῷ \ μίμνει \ κινούμενος \ οὐδέν$$
$$οὐδὲ \ μετέρχεσθαί \ μιν \ ἐπιτρέπει \ ἄλλοτε \ ἄλλῃ$$
$$ἀλλ' \ ἀπάνευθε \ πόνοιο \ νόου \ φρενὶ \ πάντα \ κραδαίνει \ [12].$$

Such a view of God as pure mind leads us to regard God as non-spatial, because minds are not necessarily but only contingently located in space. It seems reasonable to say with Boethius that God is not in space but space is present to, or (with Newton) in, God. But although we may in this sense say that God is non-spatial, we cannot analogously argue that he is non-temporal, since minds although only contingently located in space are necessarily "located" in time (see § 2, pp. 7–8; § 7, p. 37). God's possession of space can be "*tota simul et perfecta*", because it can all be present to Him spacelessly, present in His mind: but God's possession of time cannot be *simul*, though it may be *tota* and *perfecta*, and it cannot be timeless. Present it may be in some senses, but not as a temporal analogue of the Sensorium.

It is not only the spatial metaphor that is misleading St Augustine and Boethius. In seeking a God's-eye view of time, they seek to take away from time all human imperfections, and to achieve an entirely non-egocentric account. St Augustine and Boethius seek to purge the temporal language of God of all token-reflexive terms, or rather, of all those that seem to imply limitation. Quite apart from any spatial metaphor, all time must be present to God, because none can be absent.

Quod igitur interminabilis vitae plenitudinem totam pariter comprehendit ac possidet, cui neque futuri quidquam ABSIT, nec praeteriti fluxerit, id aeternum

[11]   Ch. IV, 1, 59.
[12]   Xenophanes *apud* Simplicius, in *Aristotelis Physicorum Commentaria*, 23, 11 + 23, 20; reprinted in G. S. Kirk and J. E. Raven, *The Presocratic Philosophers* (Cambridge, 1957), No. 174; and H. Diels, *Die Fragmente der Vorsocratiker*, Fr. 26 + 25: "Always he remains in the same place, moving not at all; nor is it fitting for him to go to different places at different times, but without toil he shakes all things by the thought of his mind."

esse jure perhibetur, idque necesse est et sui compos PRAESENS sibi semper adsistere et infinitatem mobilis temporis habere PRAESENTEM. [13]

Sed praecedis omnia praeterita celsitudine semper PRAESENTIS aeternitatis, et superas omnia futura... . Anni tui dies unus, et dies tuus non cotidie, sed HODIE. [14]

Nam quidquid vivit in tempore id praesens a praeteritis in futura procedit nihilque est in tempore constitutum quod totum vitae suae spatium pariter possit amplecti. Sed crastinum quidem nondum adprehendit, hesternum vero iam perdidit; in hodierna quoque vita non amplius vivitis quam in illo mobili transitorioque momento. [15]

But from this, granted St Augustine's confusion between instants and intervals (see §§ 3–4, pp. 17–23), it is a short step to another doctrine of the ever-present present, which makes time static and non-temporal. *"Anni tui omnes simul stant, quoniam stant; nec euntes a venientibus excluduntur, quia non transeunt: isti autem nostri omnes erunt, cum omnes non erunt. Anni tui dies unus; et dies tuus non quotidie, sed hodie, quia hodiernus tuus non cedit crastino; neque enim succedit hesterno. Hodiernus tuus aeternitas."* [16] Boethius makes the argument quite explicit. *"Nostrum 'nunc' quasi currens tempus facit et sempiternitatem, divinum vero 'nunc' permanens neque movens sese atque consistens aeternitatem facit."* [17] But this is to confuse the present interval with the present instant. It may be true that the interval that God can regard as "the present" is different from ours; but the present instant must be the same for God as for us, else He would not be in time at all, and so could not be a person. God's instantaneous '*nunc*', like ours, *"quasi currens, tempus facit"*. The instantaneous present is always moving on, and can be said to "make time" in the way that a moving point is said to "make" a curve. Time is the set of all

[13] *De Consolatione Philosophiae*, V, ch. VI, ll. 25–31: "That then which comprehends and possesses the whole fullness of an endless life together, to which neither any part to come is absent, nor of that which is past has escaped, is worthy to be accounted everlasting, and this is necessary, that being no possession in itself, it may always be present to itself, and have an infinity of movable time present to it."
[14] *Confessions*, XI, ch. XIII, xvi: "But thou comest before all past times by the height of thy ever-present eternity, and thou goest beyond all future times ... Thy years are but one day and thy day is not merely a day but today."
[15] *De Consolatione Philosophiae*, V, ch. VI, ll. 12–18: "For whatsoever lives in time, that being present proceeds from times past to times to come, and there is nothing placed in time which can embrace all the space of its life at once. But it has not yet attained tomorrow and has lost yesterday. And you live no more in this day's life than in that movable and transitory moment."
[16] *Confessions*, bk XI, ch. XIII, xvi: "Thy years stand all at once just because they *stand*. They do not go, being pushed out by others coming in their place, because they do not move at all: but our years shall all be when they all are no more. Thy years are but one day, and thy day is not merely a day but today, for thy today does not give way to the morrow, nor does it succeed a yesterday. Thy today is eternity."
[17] *De Trinitate*, ch. 4, ll. 72–4: "Our 'now' as it were runs along and makes time and everlastingness, but God's 'now' is indeed permanent and motionless and stable, and constitutes eternity."

instants to which the word 'now' may be (in a tenseless sense of this mood) applied, just as a curve is the set of all the points described by – or as – the moving point. Boethius has transmuted a real contrast between God's present interval and our present interval into an impossible one between his instantaneous '*nunc*' and ours, and argues that because God's present interval is infinite, therefore His present instant is static.

Boethius and St Augustine can say God's present interval is infinite. "*Aeternitas igitur est interminabilis vitae tota simul et perfecta possessio.*" [18] "*Praecedis omnia praeterita celsitudine semper praesentis aeternitatis; et superas omnia futura, quia illa futura sunt, et cum venerint, praeterita erunt.*" [19]. This much they can say. For God the present age includes all time. There is no future that has not yet come over His horizon, no past that is lost beyond recall. He is a spectator of all time, in a way in which we are not. We have our little hour, our little day, our brief life, but then we are gone, and are as though we had never been. Our present is limited: an interval, but only a finite interval. God's present is an interval too, but an infinite one, one that includes all time, past time before, and future time after, the extremest bounds of any interval which we in any context could call present.

There is no objection to saying that the interval that God regards as present is longer than any interval any man regards as present.

> A thousand ages in Thy sight
> Are but an evening gone.
>
> *Anni tui dies unus.*

Nor, although this is less intuitively obvious, is there any mathematical objection to saying that God's present is infinite where man's is only finite. It is only to be expected that an infinite God should have an infinite present. The intervals that we mortals regard as present seem often, although not always, fleeting and transitory: the present day, the present term, the present year, our present lives, seem often to be fleeting, transitory, slipping past us before we can lay hands upon them, and coming to their inevitable end. But the present century, the present era, and the present world (contrasted with the world to come) seem fairly permanent, in perhaps something of the way that God's present always is. We can reasonably discount the few human contexts in which the word 'present' is relatively stable in its reference, and can say that it is characteristic of the human condition that the present is transient. Not so with God.

To understand eternity therefore we should not think of it as timeless or changeless, but as free from all those imperfections that make the passage of time for us a matter for regret. Eternity is not temporal because with God good

[18] *De Consolatione Philosophiae*, V, ch. VI, ll. 9–11: "Eternity, therefore, is complete, simultaneous and perfect possession of unending life."
[19] *Confessions*, bk XI, ch. XIII, xvi.

things do not all have to come to an end. God, unlike Agathon (see § 8, p. 43), does not deplore the unalterability of the past, because God, unlike men, has never done anything which He afterwards could have wished He had done differently. God, unlike men, does not feel the future bearing down on Him and pressing upon Him, because He can make all His dispositions in good time, and is not going to be caught unawares, and does not feel caught up in a rush and in need of extra time to take stock and think things out properly. He feels no call ever to say

> Time you old highwayman, will you not stay,
> Will you not linger just for one day.

All our feelings about time are coloured by our own imperfections: our limited lives, the impermanence of what we love and value, the weakness and fallibility of our judgement, our failures of will, our inability to cope with circumstances or to control the course of events, our unachievements and our straight sins. Being subject to all these we think we are subject to them because we are subject to time – because, that is, we are sentient, conscious, and occasionally rational agents. And so in a sense we are, but often not in the sense we say we are;

> Time like an ever rolling stream
> Bears all of us – his sons, his weak, ineffective sons – away;

but time does not bear God away, because God is not weak or ineffective, which is as much as to say that God is not one of time's sons, but rather is the father of time, to whom we rightly say

> *praecedis omnia praeterita celsitudine semper praesentis aeternitatis, et*
> *superas omnia futura ...*

God is the master of events, not their prisoner; time passes, but does not press. All time is present in the divine mind, in the sense that none is remote or far away or absent, but not in the sense that all is simultaneous, nor that eternity is a timelessness in which nothing ever happens nor can be conceived of as happening. On the contrary

> *Illic ex sabbato succedit sabbatum*
> *Perpes laetitia sabbatizantium*
> *Nec ineffabiles cessabunt jubili*
> *Quos decantabimus et nos et angeli.* [20]

[20]   Peter Abelard:

> There each festival is followed by a further one;
> No term is set to the happiness of the holiday-makers.
> Nor shall we cease those songs that defy description,
> Which we and the angels together shall sing.

§ 56
Alpha and omega

St Augustine is led to consider time by raising the question of creation, and what God was doing before He created the universe. We still are puzzled about the beginning of time, and even more so, with a more personal concern, about its end. Did time have a beginning? Will all things come to an end? If so, what happens next?

The way we think about time is conditioned by the way we measure it. We measure it by imposing on it a discrete set of instants, marking off intervals deemed to be isochronous. The set of instants had order-type $\omega^*+\omega$ (the order-type of the negative integers followed by zero followed by the positive integers), we said (§ 12, p. 70), stretching to infinity in either direction, with no beginning, no ending, and every point exactly like every other. But perhaps we were wrong to ascribe order-type $\omega^*+\omega$; perhaps we should have ascribed to it only the order-type ω (the order-type of (zero and) the positive integers only), which does have a beginning although it has no end. This would correspond equally with what we know to be the case; indeed, it would accord better inasmuch as each one of us in his own private experience of time knows that it had a beginning although he does not know it will have an end.

Either is possible. From our point of view, which, if there was a beginning of time, is a very long time after it, there is no decisive difference between the order-type $\omega^*+\omega$ and the order-type ω. We might plead on behalf of the former that it exemplifies the principle of origin-indifference more perfectly; but the arguments that led us to formulate that demand require only the latter. We cannot claim that time could not have had a beginning on the grounds that, if it had, causal laws and communication would be impossible now.

We are more often concerned with the claim not that time, but that the universe, had a beginning. Not only theologians, but many cosmologists, think it had. Some have argued that since entropy is always increasing, it must once have assumed its minimum value. Some have argued from the red-shift and the apparent recession of the galaxies that, extrapolating backwards, there must have been a date when the whole universe was concentrated in one dense atom at one point. Some have argued that not all natural phenomena keep in step, and that some require one time scale – the t-scale – if they are to be date-indifferent, while others require another – the τ-scale (see § 14, p. 81). We then find that if one set of phenomena is to be date-indifferent, the other gives a natural origin (not necessarily in the past: it could be in the future). The last argument cannot be coercive. It is always possible to prefer some other time scale. By the regraduation theorem of § 13, we can always regraduate our measure of time so as to assign an infinite magnitude to any particular duration, and thus to relegate the putative date of creation to $-\infty$. It is a topological rather than a metrical question: whether the temporal interval is closed towards its earlier boundary – in which case there is a first instant – or whether it is open – in which case there is none. Some cosmological considerations could, granted certain assumptions, weigh in favour of one or the other alternative.

I shall not enter into an exact assessment of these arguments. Sufficient that they are respectable. They could be valid, their conclusion could be true. If so, further questions would raise. Should we describe the creation of the universe as the beginning of time, or only the beginning of change? Theologians have tended to say the former, "*Omnia tempora tu fecisti, et ante omnia tempora tu es.*" "*Procul dubio non est mundus factus in tempore, sed cum tempore*" [1]. But this is to ascribe to God a timeless, and therefore impersonal, pre-existence. If God exists from everlasting, then so must time; not simply because it is incoherent to say '*ante omnia tempora*', but because time is a concomitant of consciousness, and therefore of God. If the ultimate reality is personal, as theologians are committed to asserting, then time is eternal, in the sense of the

[1]   St Augustine, *Confessions*, bk XI, ch. XIII, § xvi; *De Civitate Dei*, bk XI, ch. VI. But it is not the only view. Gassendi denied it: "*Et, dum quaeritur, ubi esset Deus, antequam Mundum conderet, non negari quidem quin foret in se; sed concedendum simul esse, fuisse eum ubique, seu in omni loco, hoc est non in eo solam spatio, in quo Mundus erat futurus, sed in caeteris etiam infinitis.*" (Pierre Gassendi, *Syntag. Philosoph.*, pt II (Physica), sect. I, lib. I, ch. II; reprinted in *Opera Omnia* (Lyons, 1658), vol. I, p. 19). And so did Isaac Barrow and the Cambridge Platonists: "*Ita prius mundo, et simul cum mundo (licet extra mundum) tempus fuit, et est.*" (Isaac Barrow, *Lectiones Geometricae* (London, 1760); reprinted in *Mathematical Works of Isaac Barrow, D.D.*, ed. W. Whewell (Cambridge, 1860), II, p. 161). See also Henry More's letter to Descartes, quoted on p. 12, n. 12, above.

last section. Time is not a substance, something like other things in the universe, something that God could have created along with other created things. Time could not have been created any more than absolute space could be moved (see § 30, p. 142). The only sense in which God can be said to have made it, is the sense that His existence implies the existence of time. He made time only in the way in which love made Him suffer.

But if we say that the creation of the universe was the beginning of change rather than the beginning of time, we are faced with other difficulties. We may ask, with Leibniz, first, why it did not happen a year – or some millions of years – sooner; and secondly, whether it makes any sense even to ask such questions [2]. We cannot answer the first question; but, not being committed to Leibniz's principle of sufficient reason, we need not be embarrassed by our inability. Not every question can be answered. In all probability, we could not say why the initial condition of the universe was what it was, supposing we knew what it was; and neither need we explain why its date was what it was, again supposing we knew it. More serious is Leibniz's second question. What does it mean to suggest that the universe might have been created a year sooner? How could we have told? We could not have had watches in advance of creation to tell the time when it actually took place. If time is the dimension of change, then when there was no change there was no time either.

But Aristotle was wrong. As we have seen (§ 2, pp. 8–13), time is not just the dimension of change, although we rely on change to furnish us with a measure of time. Even if there were no external changes – even if we had not finished off the Amorites while the sun and moon stood still (see § 10, p. 63) – the stream of consciousness by itself would be enough to prove the passage of time. If one man had many thoughts in a flash of public time, we would say he was a quick thinker, or inspired, or something like that; but if everybody did, we should be right to say that the clocks had stood still. Time is not the dimension of change but an extension of mind, and therefore we can conceive its passage even in the absence of all change. This is why we find creation puzzling. Although *ex hypothesi* we should not have been there, we conceive ourselves being there. We take, as so often in philosophy, a God's-eye view. We consider what it would be like for God to have created the world and to have witnessed the creation of his handiwork. It is an intelligible enquiry. It gives sense to the concept of time without change, and makes us feel dissatisfied with Leibniz's and Aristotle's claim that the concept is impossible [3]. But it leaves their questions unanswered.

The difficulty lies in our differing approaches to time – personal time as the concomitant of consciousness and as the condition of choice, and public time as the dimension of change. Time as a concomitant of consciousness, a

[2]   Third Letter to Clarke, § 6; Fourth Letter to Clarke, § 15; in H. G. Alexander (ed.), *The Leibniz-Clarke Correspondence* (Manchester, 1956), pp. 26–7, 38–9. See, earlier, St Augustine, *De Civitate Dei*, bk XI, ch. V.
[3]   See also F. H. Brabant, *Time and Eternity in Christian Thought* (London, 1937), p. 22.

distentio animi, can exist without change in the external world, and is experienced by each conscious being as a magnitude, as having order-type θ. Time as an aspect of change, a symptom of symptoms, has only order-type η, but we can choose a subset of instants picked out by certain periodic processes, of order-type $\omega^*+\omega$, which will define a metric. If there is consciousness without change or change without consciousness, our everyday concept of time is deprived of one of its parents. If we are theists (or philosophical egotheists) we consider (or contemplate) the passage of time before creation, and take it to be God's (or our own hypothetical) private time, with magnitude, but no way of measuring it. If we are atheists, we consider the passage of time after creation but before the emergence of any conscious organism, and take it to be public time, with many changes available to measure it, but all of them equally conventional, and with nothing to choose between one time scale and another – moreover, with a half sense of there being nothing really to measure, only an order of change, of order-type θ, perhaps only a succession of one dam' thing after another.

If we give the primacy to private time, then we can envisage time before creation, and think of it as a magnitude, but have no means of measuring it. Leibniz's question fails not because God and time did not exist before the beginning of the universe, but because there were no years then. Leibniz can ask 'Why did not God create every thing sooner?' and receive the only possible reply: 'He did not want to'. But he cannot mention any measure of time. And therefore also he cannot invoke any principle of date-indifference. For date-indifference is a principle imposed on public time in order that we may have a rational [4] measurement of it. Public time is, by stipulation, homogeneous, just as Newtonian time is, by stipulation, reversible (see §§ 14, 49). It is open to the Identity of Indiscernibles argument that Leibniz advances against time being something independent of change. But private time does not have to be homogeneous in the same way. We experience it as having magnitude, but we do not have any intuition that any one interval is much of a muchness with any other. On the contrary, all times are different to him who is experiencing them, if only because he experiences them in order, and his memories of the earlier ones colour his experiences of the later ones. In the humanities, as opposed to the natural sciences, a difference of time is *per se* a relevant difference. Times have altered, we say. And so too, if we take a God's-eye view of creation, there is no reason in principle why the time at which creation took place should not be different from other times when it might have taken place, independently of the changes engendered by creation.

The end of the world should be thought of in the same way. It may be that the universe will come to an end. If so, time will not come to an end, although we shall no longer be able to, or need to, measure time. In the absence of an

[4]   In the sense connected with 'reason', not that of rational numbers as opposed to real numbers.

external world, it will be impossible to impose a naturally given metric on time: time will have no metrical properties, only topological ones. So too in the after-life, if there is one. Time should be thought of as devoid of metrical properties in the world to come, at least of those that imply limitations. Yet this is difficult to conceive. For God eternity is possible, and would be a good: for us, limited beings, if not mortal, eternity is not so obviously possible, nor so obviously a good. We might be bored. Boredom, according to Plotinus, was the reason why time had to exist; but boredom for many people now is the reason why they no longer want to exist, and see in everlasting life a prospect not of eternal bliss but of infinite tedium. Many people do not find life worth while. They can seek transitory pleasures in the short time that is given them before they go down to their long home, but their resources are too limited to enjoy endless existence. Only because the time for gathering them is scarce do rosebuds seem worth having: if they had to be considered for their value in themselves, apart from any scarcity value, many would find them not worth the effort of picking up. Not only do pleasures pall, but they would not be felt to be fun even at first if they were not known all along to be fleeting. Our lives are lived under the limitations of mortality, and many of our everyday values depend on those limitations for their being precious in our eyes.

It follows that immortality is not an easy notion, nor, if it be possible, something easy to be wished for. We have an intimation of this in the possible progress of geriatrics. If geriatrics reached the stage where people no longer died – and not only did not die, but could be preserved in, say, late middle age – we should find some who needed to be amused or occupied in order to while away the time, and others who wanted all the time they had in order to do what was worth while. Those who found time hanging heavy on their hands would not, were it not for the fear of death, welcome extra time. The test of time is a severe test of value. Only something supremely worth while can stand having infinite time to do it in. Else – if we were like those who devote their lives to killing time – infinite time would be everlasting hell.

Some people say we should forget the future and concentrate on making the most of the present.

> *quid sit futurum cras, fuge quaerere et*
> *quem Fors dierum cumque dabit, lucro*
> *appone nec dulcis amores*
> *sperne puer neque tu choreas*
>
> *donec virenti canities abest*
> *morosa.* [5]

[5]   Horace, *Odes*, I, ix, ll. 13–18: "Stop asking what will happen tomorrow, and reckon each day that fortune grants as a bonus. Dance and make love while you can, before sour old age prevents you."

sapias, vina liques, et spatio brevi
spem longam reseces. dum loquimur fugerit invida
aetas: carpe diem, quam minimum credula postero. [6]

We need not enjoy the concert any less for that we know that it must end soon;
nor is the present less valuable because the future is blank. Sufficient unto the
day is the value thereof, and we are wrong to worry about what is yet to come
when we should be content with what we actually have.

We may take the point. And indeed, it is true that if we cannot enjoy the
present, we shall not be able to enjoy the future when it comes. But the argument applies the other way too. The future, just because it will one day be
present, cannot but have some bearing on our present state of mind. We are
rational agents, making choices about what we shall do, and thus implicitly
making up our minds about which of various possible future states of affairs
is most worth bringing about. We cannot, so long as we have minds, avoid
extending them towards the future, or take no thought whatever for the morrow,
and cannot, while remaining rational agents, divorce time present from time
future and time past, and concentrate on it to the exclusion of all else. We
should not be in fetters to the future, but if we take a long view of ourselves,
it must include some indication of what is going to happen, particularly what
is going to happen at our hands, and what has happened, as well as what is
happening now. We can enjoy the present for what it is worth, but what it is
worth depends on the context in which it is to be seen. Horace's youth can woo
his maiden happily, believing the affections of his heart to be the real thing; but
the dalliance of the middle-aged man, undertaken in the knowledge that love
does not last, soon palls. One can enjoy one holiday, putting aside workaday
worries, because one knows it is a holiday, a period of recreation, in a life that
as a whole is useful and creative. But those very same activities, fully enjoyed
though they now are, would become insupportably tedious if they were all
that one did, or were set in the context of a life that was not, as a whole, worth
while – as many people who have retired on account of fortune, health or age
have discovered to their cost. Our activities, although not completely inter-
locked, are not completely separate, and they affect one another in the value
we can place on them; and in particular, the value of present activities, although
real, depends on some conditions about the past and future being fulfilled. It is
difficult to enjoy gardening if tomorrow the garden is to be destroyed by the
bulldozers, just as it is difficult to enjoy the prize at the flower show if one's
flowers were not really grown by oneself. When the atomic bomb was discovered, many people believed, with reason, that the world would soon be
destroyed, and found that that expectation took away the value of all that they
did, and made everything seem futile and pointless. And the same applies to

[6]   ibid. xi. ll. 6–8: "Be sensible. Cut out ambitious plans, decant the wine, for we have but
a brief space, and even while we are speaking jealous time has fled. Gather the fruit of the
day, relying as little as possible on what the future may bring forth."

each individual as he contemplates his own future, and the implications of his own mortality.

For we are afraid of death. *Timor mortis conturbat me.* It is difficult therefore to be entirely dispassionate about the passage of time.

> *Pallida Mors aequo pulsat pede pauperum tabernas*
> *regumque turris. O beate Sesti,*
> *vitae summa brevis spem nos vetat incohare longam.*
> *iam te premet nox fabulaeque Manes*
> *et domus exilis Plutonia.* [7]

And so we come to regard time not, as we should, as the passage from aspiration to achievement, but as a relentless process carrying us on to doom and destruction. Yet if there is no God, then death cannot be an absolute evil, since immortality, without God, would be insupportable. And if there is a God, then death, although an evil in many respects, cannot be an absolute one for those who even in this life have come to set their hearts on Him. We do not know. In an important sense, we cannot know. We have intimations of both. We have moments of inspiration and of desolation, periods of high endeavour and dreary stretches when nothing seems worth doing and all appears grey. We know in advance what death – at least the death of the spirit – must be like; and there are times when we are permitted ἐφ' ὅσον ἐνδέχεται ἀθανατίζειν, and know what immortality would be like and how we could want to have it. But in our ordinary, untransfigured moods, our search for ultimate truth or the end of existence is, as it must be, a search without certitude, in which the humanist can only hope, and the Christian pray, that death when it comes, should come no longer as an enemy, though not yet as a friend.

[7]   ibid. iv, ll. 13–17: "Pale Death with equal foot taps on the cottages of the poor and the towers of kings. Life is too short, my good Sestius, to allow long-term hopes. Night is coming down on you, together with the Shades and the hollow halls of Hades."

Who would have thought my shrivelled heart
Could have recovered greeness? It was gone
Quite underground: as flowers depart
To see their mother root, when they have blown,
    Where they together,
    All the hard weather,
Dead to the world, keep house unknown.

These are thy wonders, Lord of Power,
Killing and quickening, bringing down to hell
And up to heaven in an hour;
Making a chiming of a passing bell.
    We said amiss
    This or that is:
Thy word is all, if we could spell.

And now in age I bud again;
After so many deaths I live and write;
I once more smell the dew and rain
And relish versing; O my only Light!
    It cannot be
    That I am he
On whom thy tempests fell all night.

George Herbert

Index

(Where page numbers are not in numerical order, those given first are main references.)